Secret Agents Present

Looking Through A Glass Darkly

Ken Stange

BOOKS BY

KEN STANGE

A Smoother Pebble, A Prettier Shell (Penumbra Press)

Advice To Travellers (Penumbra Press)

Bourgeois Pleasures (Quarry Press)

Bushed (York Publishing)

Cold Pigging Poetics (York Publishing)

Colonization Of a Cold Planet (Two Cultures Press)

Embracing The Moon: 25 Little Worlds (Two Cultures Press)

Explaining Canada: A Primer For Yanks (Two Cultures Press)

God When He's Drunk (Two Cultures Press)

Going Home (Two Cultures Press)

Love Is A Grave (Nebula Press)

More Than Ample (Two Cultures Press)

Nocturnal Rhythms (Penumbra Press)

The Sad Science Of Love (Two Cultures Press)

Secret Agents Past: The Parting Of The Waters (Two Cultures Press)

These Proses A Problem Or Two (Two Cultures Press)

Secret Agents Present

Looking Through A Glass Darkly

~~

Ken Stange

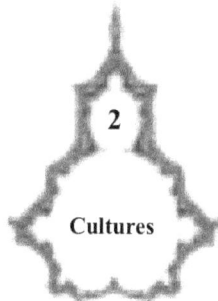

Two Cultures Press

2014

For information about permission to reprint, record, or perform sections of this book, write to **Two Cultures Press**, 970 Copeland, North Bay, Ontario, Canada, P1B 3E4 (**info@twoculturespress.com**)

Library and Archives Canada Cataloguing in Publication

Stange, Ken, 1946-, author
 Secret agents present : looking through a glass darkly / Ken Stange. -- 1st edition.

Includes bibliographical references.
ISBN 978-0-9939201-1-0 (pbk.)

 1. Creative ability. 2. Art and science. 3. Artists--Biography. 4. Scientists--Biography. I. Title. II. Title: Looking through a glass darkly.

BF408.S727 2014   153.3'5   C2014-906641-4

Acknowledgements

"Redefining Creativity: How Science and Technology Make Us Rethink Creativity" (based on some ideas in this book) was presented and broadcast as a TEDx lecture sponsored by Nipissing University. (2012)

Some the ideas in the section "Dangerous Seductions" were the basis for a presentation on "Using Interactive Teaching Of Probability To Explain Irrational Beliefs" at the International Conference on the Teaching Of Psychology (2009).

Cover design and art: Ken Stange

ISBN: 978-0-9939201-1-0

This book is dedicated to my not-so-secret agent: Ursula. Her collaboration in all my creative efforts is a secret all over the block.

Contents

FOREWORD: FOREWARNING 8

PREFACE: CAVEAT LECTOR 9

PROLOGUE: OUR IMPERFECT VISION 11

THROUGH A GLASS DARKLY (OUR ALL TOO HUMAN LIMITATIONS).. 13

THROUGH A GLASS DARKLY.. 14
COLLABORATION AND COLLUSION 16
NURTURING NATURE.. 17
AXIOLOGY AND EPISTEMOLOGY: THE AMORALITY OF CREATIVITY 22
A & E (ART & ETHICS): HEAD HUNTERS DINING OUT AND FINE ART... 23
S & M (SCIENCE & MORALITY): PHYSICISTS PLAYING AND ETHICS.. 27
DECADENT ARTISTS AND MAD SCIENTISTS 33
CASE STUDIES: VICTOR FRANKENSTEIN, HENRY MILLER 41
MADNESS AND GENIUS: ALLIES OR ENEMIES?..................47
CASTLES IN THE SKY: NEUROSIS AND PSYCHOSIS............. 48
WHAT GOES UP MUST COME DOWN: MANIC-DEPRESSIVES 50
OUT, OUT DAMN SPOT: OBSESSIVE-COMPULSIVES 53
THE CREATIVE ACT: SYMPTOM OR SIDE EFFECT OR ACT 55
THE CREATIVE ACT: THERAPY OR RETREAT OR CURSE? 57
CASE STUDIES: LORD BYRON, AUGUSTA ADA BYRON....... 61
OUR SENSORY CENSORS: OUR ALL TOO HUMAN LIMITS 67
ARTISTIC VISIONS AND ILLUSIONS 69
SCIENTIFIC VISIONS AND DELUSIONS............................... 72
MIXING OUR ALREADY DISORDERED SENSES: SYNESTHESIA.. 76
MIXING OUR MANY MANGLED METAPHORS: MULTI-MEDIA ART.. 79
BEYOND CULTURAL METAPHOR AND SENSORY LIMITATIONS: MULTI-MEDIA SCIENCE 89
CASE STUDIES: STEPHEN HAWKING, LUDWIG VAN BEETHOVEN... 92
MEMORY, LANGUAGE, AND THOUGHT: OUR UNMETRICAL MENTAL MAPS 98

THE CASE FOR FORGETTING: ROLE OF MEMORY IN THE CREATIVE METHOD.. 99

A LEGEND FOR OUR MENTAL MAPS: LANGUAGE, COGNITION AND CREATION... 105

PRE-COGNITION: MERCATOR PROJECTIONS, ZEN KOANS, AND VISUAL ANALOGIES... 110

THE TRAPS OF INTROSPECTION 113

THE PITFALLS OF ASSUMED OBJECTIVITY......................... 115

CASE STUDIES: MARCEL PROUST, GEORG CANTOR......... 119

LONGING FOR THE INFINITE: LOOKING FOR GOD AND HIS CRONIES 124

SPIRITUAL NEEDS AND THE TELEOLOGICAL URGE...... 125

GOD AS HELPER AND THE AMUSING MUSE MYTH......... 130

CASE STUDIES: ALBERT EINSTEIN, DANTE ALIGHIERI. 135

THE CONSERVATIVE URGE: FADING IMAGES AND SHIFTING PARADIGMS 141

MORE THAN A NICKEL'S WORTH: BUDDY, CAN YOU SPARE PARADIGMS?.. 143

CONSERVATIVE IMPULSES AS A CATALYST TO REVOLUTION 152

THE SYNCRONICITY OF RADICAL CHANGES IN PHYSICS AND IN ART... 155

CASE STUDIES: BLAISE PASCAL, EZRA POUND................ 165

DANGEROUS SEDUCTIONS: THE PSYCHIC CIRCUS 174

THE FALSE MYSTERIES: THE BANALITY OF THE SUPERNATURAL.. 175

THE NEW AGE OR NEW DARK AGE?................................... 188

CASE STUDIES: JAMES RANDI, ALFRED RUSSEL WALLACE 191

EPILOGUE: IMPROVING OUR VISION 198

AFTERWORDS 199

ACKNOWLEDGEMENTS... 200

SELECTED ANNOTATED BIBLIOGRAPHY........................... 201

AUTHOR'S NOTE... 212

ABOUT THE AUTHOR 212

FOREWORD: FOREWARNING

This book is the second in a trilogy on the nature of creativity in the arts and sciences. I eventually adopted a quasi-chronological approach because organizing my many scattered thoughts on this topic wasn't easy. This second book is about the present state of affairs, which has changed substantially given the increase in resources for the creative. We also know more about the nature of creativity, although what we don't know is still far greater.

The third and final book consists of speculations on the probable future, even though I know too well that predicting the future is a mug's game.

The three *Secret Agents* books:
  Secret Agents Past: The Parting Of The Waters
  Secret Agents Present: Looking Through A Glass Darkly
  Secret Agents Future: Going Where There Be Dragons

PREFACE: CAVEAT LECTOR

This book is (as are the other two *Secret Agents* books) intended for the general reader and not as a scholarly review of 'the literature' on creativity. It is intended as an informal, speculative quasi-philosophical exploration of the nature of creativity in art and in science, spiced up with a few polemics. For this reason, and for the sake of readability, I have kept the customary, often cumbersome, paraphernalia of formal scholarship to a minimum; so the reader will not find any of those intrusive APA format parentheses with the names and dates of research papers. Nor have I banned the use of the first person singular and personal anecdote from my prose, as is required in scientific publications trying to maintain an objective tone. I openly admit to being opinionated and less than coolly objective, for I care passionately about the topic of this book. For the same reason, I've tried to avoid the "Hey, I'm not responsible" passive voice.

.

Consequently the reader won't find extensive, formally constructed references to support all of the factual information included in this book. However most of what I say, not opine, can be easily confirmed independently, given access to The Internet and any decent library. In those cases where some fact or reference is not so easily confirmed by the average reader in our wonderful electronic age, or where I felt it worthwhile to point to interesting material relevant to the topic at hand, I have inserted an adequate bibliographical pointer in a footnote. In most cases I am sure that simply giving the author and the title of a reference these days is sufficient to easily locate full bibliographical information on the Internet.

.

Thus the policy I have adopted is to use footnotes primarily as a place for extended parenthetical information or comment, the occasional wisecrack that I couldn't resist but didn't want to interrupt the flow of the main text, and brief author/title pointers to relevant books or articles. Only when directly quoting a writer or when referring to extremely specific material—such as a particular research study—have I felt it necessary to include a detailed bibliographical reference in the footnote.

.

Naturally, I accept full responsibility for any errors of fact or interpretation of fact. That I've not called expert witnesses to substantiate all my statements should not be interpreted as my not checking my information. Those who wish to question something I've passed off as factual should find it easy to check up on me, and I'm more than willing to stand corrected. I'm sure there are many experts in the various fields where I've trespassed that are queuing up to do just that—and shoot me down. I only ask they don't shoot to kill.

So although this book has no references page, I couldn't resist appending recommended further reading: I have included a somewhat eccentric, quite eclectic, selected bibliography, sorted by chapter topic. Many of these works have been a source of both factual information and inspiration to speculation.

PROLOGUE: OUR IMPERFECT VISION

*"For now we see through a glass darkly; but then face to face: now I know in part; but then shall I know even as also I am known"**
—The Apostle Paul *(First Corinthians: Chapter 13; Verse 12)*

.

"Appreciation doesn't require understanding, but it does enhance it."
—Heraclitus

How can we explain human creativity? We can't. But nor can we explain the creation of the universe.

.

This doesn't mean we know nothing and can't know more. We have partial understanding and that is to our credit.

.

We know a lot about the current state of our universe, and the evidence is pretty strong that our universe came into existence 13.8 billion years ago when a 'singularity' exploded in what is called "the Big Bang". Why it exploded we don't know.

.

In the case of human creativity, the evidence is that it came into existence because it had survival value, which is what evolutionary natural selection is all about. But we don't know that much about its characteristics. However, as with our knowledge about the universe, we do know more than we did in the past.

.

Everything we have learned about both the universe and creativity should fill us with wonder, but that wonder includes much unsettling knowledge. Both our existence and our creativity are fragile.

.

Our creativity has made possible our great accomplishments: science and art. But we are all too human, and humans are imperfect. Creativity has limits. It may indeed be a gift, but it is gift that is not easily unwrapped.

.

---

* Biblical scholars point out that "glass" more likely means mirror than window, but both interpretations are apt for my purposes.

This book is a modest attempt to examine these limitations as best as is possible given our imperfect vision.

THROUGH A GLASS DARKLY
(OUR ALL TOO HUMAN LIMITATIONS)

THE PRESENT. *The central pane in the triptych: an examination of some of the refractory and reflective properties of our window to the world of creative possibilities—including a brief foray into ethics—and an examination of both popular and esoteric ideas about the creative process. It is also a close, hard look at inherent limitations in the human condition and how confrontation with our unacknowledged collaborators redefines creativity.*

THROUGH A GLASS DARKLY

"We can never cease to be ourselves."
—Joseph Conrad (*The Secret Agent*)

"He who has imagination without learning has wings and no feet."
—Joseph Jourber (*Pensees*)

Imagination often fails us, and most often because what we call innovation is limited to recombination. Whenever we try to build something new, we have to use the materials at hand, and for human beings those materials are the stuff collected from our experience. We build the imaginary unicorn from the horse and the rhinoceros. It is impossible to truly imagine anything completely foreign to our experience. This is why creativity is grounded in metaphor: the connection between the newly imagined and the previously experienced.

Whenever we try to build something new, we have to use the materials at hand, and for human beings those materials are the stuff collected from our experience, *and limited by what we are humanly capable of experiencing.* We can never really imagine what colours a bee sees, with its optical receptors for ultraviolet light, when it buzzes about our garden. Or what sounds bats can hear, which are several magnitude above our upper frequency range of hearing. If there were bees who created visual art or bats who wrote symphonies, we perceptually-challenged humans could never hope to respond to them.

But if we have to work with what we have, we will work most effectively if we understand our materials. This understanding has always been sensate, artistic, and scientific. Traditionally, science has been at the service of the artistic, dating back at least to the ancients studying mathematics as guidelines to the creation of sculpture, architecture and music. And traditionally, aesthetic concerns have guided (albeit occasionally misguided) science in its efforts. Science and art have always been collaborators in the creative enterprise, but now both seem hesitant to admit their collaboration. It is time to admit and examine this relationship.

COLLABORATION AND COLLUSION

This trilogy is entitled "Secret Agents" because the collaboration of science and art has come to be seen, if it is even admitted to exist, as more a secret collusion than collaboration. Science has done much to illuminate such matters central to creative activity as memory, personality, sensory systems, neural physiology, innate cognitive patterns, sexual needs, fear of mortality, spiritual drives, need for novelty, need for security. So has much of philosophy, of course, which I still think of as actually being "natural science".

Some of the central issues worth examination from both a scientific and artistic perspective are the relationship between ethics and aesthetics and scientific endeavour, the relationship of madness to creativity, the importance of the nature and limitations of human cognition and perception to any kind of understanding, the way creative goals dramatically change in response to sometimes random events, the role of spirituality and religion, and finally the problem of artist and scientist ignorantly latching on to what is least valid in each other's province.

NURTURING NATURE

Before venturing further, I have to have post a warning. As is probably already evident, I think it silly to examine creativity while wearing rose coloured glasses. They do not permit one to see clearly. If the glass is already 'darkly', it doesn't help matters to add another filter—even if it colours the world with a lovely rose tint. But one will see many people wearing these tinted spectacles at conferences on creativity or even at any gathering of social scientists.

.

Stephen Pinker, somewhere in his refreshing and insightful book, *The Blank Slate*, remarks that social science "isn't for sissies". He doesn't mean the psychology and sociology departments aren't overburdened with sissies.* He means that doing real social science involves uncovering some unpleasant facts that will enrage people, some of whom will come pounding on your door, screaming epithets at you. Your ivory tower is not secure.

.

In his book Pinker cites numerous appalling examples of this behaviour, especially in the 'radicalized' seventies. But of course, as he acknowledges, it still goes on. Some research topics are very dangerous territory—especially if they challenge feel-good dogmas, be they political or religious.

.

Pinker cites three dubious, but widely accepted, philosophical tenets that have proven very dangerous to challenge. They are what he ironically labels as the current liberal "Holy Trinity": the "Blank Slate"; the "Noble Savage", and "The Ghost In The Machine". The blank slate of course refers to the idea attributed to the philosopher John Locke that we are born 'blank slates' on which environment writes our future.† The conception of the 'noble savage' is usually attributed to Jean Jacques Rousseau who thought while we are all born free, we are everywhere in chains—because of civilization. And 'the ghost in the machine' refers to the mind-body dualism of Rene Descartes, which haunts the house of philosophy even today.

.

---

* I think they are, alas.

† Locke's position contrasts with Kant's idea of inherent, *a priori*, mental structures. There is nothing in Locke's writings suggesting he would have endorsed the extremist behaviorist John Watson's claim that he could make of any newborn into "a doctor, lawyer, artist, merchant-chief, even beggar man and thief" or anything of his pleasing by simple conditioning.

Pinker does a marvellously effective job of showing how the findings of hard science and common-sense challenge these assumptions. Furthermore, and importantly, he also explains why rejecting them is not an endorsement of immorality and anarchy—but actually quite the contrary. However, I mention his book here because some of the issues directly pertain to creativity and creative individuals—if they are to be examined *sans* rosy spectacles.

Much of what I have already said, and will say, undoubtedly will be seen as sacrilegious to true believers in this secular "Holy Trinity".

The Blank Slate

We are not all creative, at least in the way I've already clearly operationally defined it. And even if we are creative, we are not equally creative. I am not Shakespeare, alas. (And to lump my neighbour who has so cleverly furnished her apartment on a very limited budget, in the same category—'creative'—as Michelangelo or Einstein or Mozart or Aristotle is rather silly.) Not only are we not all creative or equally creative, part of the reason for the difference is natural 'gifts'. It would be wonderful if every music lover could take their newborn and, by applying the right operant conditioning procedures and by exposing him to the right experiences, create a Beethoven. But both science and common-sense inform us that that is ridiculous. To believe that creativity is not at least partially a result of the roll of the genetic dice is to label every parent negligent whose children don't spend their lives doing significant art or science or philosophy. Besides, Beethoven wasn't a particularly nice, moral, or happy man. Might it not be better to have children of less musical talent, but more morality and common decency and compassion? This is a serious and difficult question.

It should be obvious that I value creativity very highly, but that does not mean that I, or anyone who stands in awe and appreciation of the highly creative, denigrate everyone else. The world would be more improved by an increase in the number of moral, decent human beings than by an increase in numbers of even the most creative. (It is an easily documented fact that these two characteristics do not correlate as strongly as those still sporting rosy glasses would like to believe.)

What is at issue here is the pseudo-egalitarianism that most people associate with the infinite human potentiality that the blank slate conception implies. We are not all born equal, but if we are born in a civilized and sensible society, we are all given equal opportunity to actualize what potential nature has given us. We devalue creativity by claiming it is universal. We devalue people by laying on them or their parents any failure to be creative—because for god's sake they had the chance! (Or so we falsely assume.)

.

How much of creativity is nature and how much nurture? One would think this old and decrepit question would have been put to its final rest by now. Why is it still one of the walking dead, rustling its ghostly chains in the halls of social science departments? Well this is explained in Stephen Pinker's book, which it makes more sense to recommend than summarize. I'll only make a few remarks.

.

The psychologist Donald Hebb allegedly responded to the question of which matters more, nature or nurture, with this apt analogical retort question: "Which matters more to the area of a rectangle, the length or the width?" Depends on the rectangle, of course. If you're born with great exceptional creative length, say the brain of a Mozart, then the width of your environment will have to be pretty damn great to have much of an effect on your area of accomplishment.* On the other hand, if you're born with a more modest creative length, then whether your environment is enriched (wide) or deprived (short) will make a large difference.

.

To phrase it in statistical terms, if you have only two variables that affect something, decreasing the variance in one will increase the importance of the other. Take extreme cases. If two creatures are exposed to *exactly* the same environment from conception, any genetic differences will account for *all* differences between them. If, on the other hand, you have two clones, even the slightest difference in the environment in which they are raised will be of monumental importance in explaining why they differ as adults. Every biologist knows this, and it is taught in Genetics 101. Many social scientists and social engineers should register for this class.

.

The Noble Savage

.

---

* An example of a 'great' environmental width in this case would never having any exposure to music. We obviously wouldn't have the Jupiter Symphony if Mozart was born in Tasmania.

No such thing. Hobbes better describes life in most hunter-gatherer cultures as "poor, nasty, brutish, and short"[*] than by Rousseau's image of the noble savage, peacefully living in a state of nature, munching on free fruit. This is a well-established fact[†], and the scandals associated with the bad feel-good anthropology of the likes of Margaret Mead remain an embarrassment to serious social scientists. That modern civilizations in the twentieth century have regressed to horrific savagery such as perpetrated by the Nazis and Fascists and Communists[‡] does not mean that civilization itself is bad—and that being a 'savage' is good. It means that civilization is a fragile social construction built on the unstable ground of our natures as animals.

.

Nature is, as Tennyson put it, "red in tooth and claw", something any serious study of zoology will confirm. Going 'back to nature' is *really* not a good idea. You wouldn't want to be a chimp or a baboon, trust me—even if you were an alpha male. Contrary to current usage, it would make a lot more sense to attach a positive connotation to 'unnatural' than to 'natural'.

.

Art is unnatural, as is science. We mark the beginning of civilization, even our species, by the first archaeological evidence of these things: i.e., art by cave painting; science by tool usage.

.

Creativity itself is unnatural. When some evidence of it is discovered in species other than ours, it is a source of great excitement. Kohler's monkey figuring out how to combine two poles to get that out-of-reach banana or Thorndike's insightful cat figuring how to open the latch to escape that damn cage ("puzzle box") are big news.[§] Wow, animals can have creative insights and solve simple problems!

.

---

[*] In his classic work *Leviathan*.

[†] For those that would question this, Stephen Pinker includes an interesting graph (on page 57 of *The Blank Slate*) of percentage of male deaths caused by warfare in eight typical hunter gather cultures compared to the U.S. and Europe in the 20th century. Hard as it is to believe, modern man is downright pacifistic by comparison. (Pinker gives the original source, for those who, understandably, that find this incredible.

[‡] The 'apogee' and nadir of social engineering—which seems to go unnoticed by current reformers.

[§] These are two famous examples of animals showing 'insight' that are almost always cited in Introductory Psyc textbooks. Wolfgang Köhler was a gestalt theorist. Edward Thorndike is one of the founders of learning theory. His "Law Of Effect" and theory of connectionism is the cornerstone of conditioning as learning.

So creativity is unnatural, or at least rare and atypical, in the natural world. There is nothing noble about being more natural, which would mean returning this wonderful gift that the vagaries of evolution have bestowed on us. One could say the human species is a 'freak of nature' in sometimes possessing this characteristic. And the corollary thus is that those humans who possess a great deal of it are truly 'freakish'.

.

The Ghost In The Machine

.

This is the thorniest of philosophical issues, but probably the least problematic in any discussion of creativity. While it is of great importance in any discussion of human accomplishment (or, for that matter, disgraceful behaviour) to know if credit (or blame) is to be laid on the individual—or whether it was inevitable in a purely deterministic universe—this question is not of central importance in trying to understand creativity.

.

However, the interaction of emotion and reason is important, especially in terms of relating scientific to artistic creativity. Also important is the relationship of brain function and our purely physical and physiological limitations to our creative endeavours. In fact this is a central theme in this middle pane in my triptych.

.

So this ghost, this assumed homunculus inside each of us, will occasionally make an appearance. It just won't be on centre stage.

.

On to our all-too-human characteristics…

AXIOLOGY AND EPISTEMOLOGY: THE AMORALITY OF CREATIVITY

"It had been a wonderful evening and what I needed now, to give it the perfect ending, was a little of the Ludwig Van... Oh bliss! Bliss and heaven! Oh, it was gorgeousness and gorgeousity made flesh. It was like a bird of rarest-spun heaven metal or like silvery wine flowing in a spaceship, gravity all nonsense now. As I slooshied, I knew such lovely pictures!" (Spoken by Alex, the thug and protagonist of the book, while listening rapturously to Beethoven's Ninth Symphony—after brutalizing several people during a night out on the town.)
—Anthony Burgess (*Clockwork Orange*)
.

"The passion for aesthetic perfection, like all passions, relegates ethical considerations to mere obstacles on the way to consummation. All is fair in love and war—and art and science."
—Hippokrites

We expect too much of creative people. Why on earth should we expect them to be any nicer than our grocer, banker, butcher or candlestick maker? It may be that they see farther or deeper—or feel farther and deeper—but why should such deeper insight automatically engender higher morals? Knowledge is power, and power, it is widely acknowledged, corrupts.
.

Good sometimes arises out of evil. Beauty sometimes arises out of ugliness. It is a nice sentiment that truth is beauty, and beauty truth, but it isn't really true. It is also not true that someone who creates something beautiful which is of scientific or artistic value is necessarily a good person.
.

Even more disturbing is the fact that much—virtually all—of the great art that moves us so deeply could never have been conceived of and created if the world were utopian.

A & E (ART & ETHICS): HEAD HUNTERS DINING OUT AND FINE ART

When in an attempt to bring civilization to the people living in the jungles of Western New Guinea, the new Indonesian government of Irian Jaya declared the eating of *daging orang* (human flesh) illegal. And art suffered a terrible loss.

Cannibalism, headhunting and relentless tribal warfare are not characteristics of civilization. And few would call 'civilized' a culture that has no science, only primitive wooden tools, and a religion based on the belief that all deaths (except of the very old or very young) are the result of either evil worked by living enemies (if not directly, then through black magic) or the action of dead spirits seeking revenge on their murderers and their murderers' tribe—and that all deaths must be avenged. The Asmat, a semi-nomadic people inhabiting the swampy coastal rainforest and tribal rivers along the southwest coast of Irian Jaya, lived lives that are by any reasonable standard nasty, brutish and short. In fact, to call theirs a "Stone Age Culture" (while accurate since they have no metal tools) may be to malign the Neanderthals and Cro-Magnons. Yet these people did possess one hallmark of civilization—a bold and refined art. They did, that is until they were 'pacified' and 'civilized' in late 1960s.

The Asmat culture* is one of those many where there is no specific term for art, but where art is inextricably entangled in religion, where 'art for art's sake' is unheard of, where art serves a 'higher' purpose. The Asmat people believe in three worlds: the world of the living; the world of active and often malevolent spirits of the dead; and the world of "ancestors", spirits finally put to rest, a place which has been compared to the Christian conception of heaven. Their 'spiritual' goal is to maintain a balance between these three worlds: this is accomplished by ritual celebrations. Their art is central to these religious celebrations; in one sense their art defines it. Beautiful war shields, body masks, drums, "ancestor" figures and poles, horns, spears, masks, canoes, paddles, and other art objects were created solely for these events—and their highly specific ties to the ritual and religious observance is evidenced by the fact that some were simply

---

* For a website dealing sympathetically with the Asmat, go to http://www.szgdocent.org/ff/f-asmat.htm

discarded as worthless after they had served their purpose in some celebration.

.

But these feasts and celebrations involved ritual mutilation of the heads of enemies, cannibalism to absorb the powers of the enemy, and wild orgies where the flow of semen was thought to rebalance the spirit world with that of the living. One can imagine the shock and horror of the first explorers and missionaries to venture into this strange world. Surely there was nothing 'noble' about these savages. (While watching human flesh being devoured and strange things being done to human heads and skulls, one isn't likely to notice the fine carvings and paintings on the totems, masks and shields that are part of the proceedings.) So in the 1960s a strict ban on ritual warfare, headhunting and cannibalism was enforced as best it could be, given the isolation of the tribes. The "men's houses" filled with the skulls of ancestors were burned to the ground.

.

This was eventually (and inevitably) followed by the invasion of the entrepreneurs, always anxious to exploit 'untapped' natural resources, including cheap labour. The Asmat were only too willing to be exploited. They loved things they would normally have no access to—especially tobacco, steel axes, matches, fishing line, and food other than the insipid balls of sago sap that was the staple of their diet. This introduction of trade, more than the prohibitions of some incomprehensible thing called government, is what eventually led to their pacification, and the end of ritual warfare and the headhunting and cannibalism associated with it. The Asmat were soon living in clapboard houses with tin roofs to protect them from the incessant rain and malarial mosquitoes—and working for the strangers with pale skin and an apparently endless supply of really cool, exotic and useful stuff. The good, the bad, and the indifferent changes associated with this familiar tale of an aboriginal people being 'civilized' are always difficult to weigh; but, in this case, only the most naïve romantic would try to deny that the lives of the Asmat were made at least a little less short, nasty and brutish. The passing of a culture based on never-ending revenge-warfare, where considered essential is not only killing, but atrocities as well, is not a passing to be mourned—even if the version of civilization with which it is replaced is less than morally exemplary. But the irony and paradox involving creativity is that it seems the beautiful baby had to be thrown out with the filthy bathwater.

.

In an essay that points to the paradoxes inherent in the idea of civilization, the intrepid travel writer Tim Cahill describes his trip to Irian Jaya in the late 1990s.* Desirous of purchasing some Asmat art, he is taken by Rudy, a local 'art dealer', to a clapboard shop and offered crudely chiselled pornographic carvings—"Sexy", Rudy whispers leeringly—which bear as much resemblance to the real art of the Asmat as those $29.95 "original oil paintings" of flowers sold by street vendors in Paris do to the work of a Monet or a Matisse. Cahill then ventures deep into the interior to a small Karowai (or Korowai) village, where he discovers in this neighbouring, and still relatively unchanged, tribe something of the way life once was for the Asmat—but does not return with any art, only a sense of the contradictions of so-called civilization.

.

Original Asmat art, the art created before the intervention of the outside world, is highly valued by connoisseurs of aboriginal art. Following the end of World War II, various missionaries began proselytizing in the remote areas of New Guinea. By 1953 the Catholic Crosier Order had become established in the small town of Agats. This order has been credited with saving much of the art of the Asmat people, as well as being a more benign cultural influence than is usual for missionaries. In the early sixties the Harvard Peabody expedition set out to purchase Asmat art for an exhibition in the United States. The subsequent disappearance, and presumed consumption, of art collector Michael Rockefeller led to a tabloid feeding frenzy about the actual feeding habits of New Guinea natives. Shortly thereafter the newly empowered Indonesian government responded by a serious crackdown on traditional native practices, combined with severe restrictions on foreign visitation. But the 1990s saw the restrictions lifted and a renewed interest in Asmat art. There are now substantial permanent collections at the Metropolitan Museum of Art in New York, the Peabody Essex Museum in Salem, Massachusetts, the Sowada Gallery in St. Paul, Minnesota, and the Utah Museum of Fine Arts. At least one expert in the field, Tobias Schneebaum, claims that Asmat art is undergoing a Renaissance, but the fact remains that the original art was part and parcel of the whole culture—a culture whose values were morally abhorrent—and when the culture changed, the art it generated ended.

.

What we have here is a dramatic, even melodramatic, example of the paradox of creativity in the service of cultural values. It is darkly

---

* "Among the Karowai: A Stone Age Idyll" in *Pass The Butterworms: Remote Journeys Oddly Rendered* Vintage Books, 1998.

amusing to try to fit this into the framework of the goodie-two-shoes creativity theories of psychologists such as Maslow, with their hierarchy of values: The Asmat achieved the highest levels of "self-actualization" through chopping off the heads of those not of their tribe, eating their tasty bits, and creating fine art to celebrate and affirm this achievement!

It is undoubtedly true that creative individuals flee repressive regimes, and that artists usually seek out a place to work where the government or the predominant culture is not dangerously hostile to them and their work. But they usually do so, not out of some elevated morality, but rather just so they can (damn it!) get on with work. Where their work is not threatened by the prevailing cultural or political climate (or maybe even approved of), they just go on with their business—and sometimes create things of lasting values. Leni Riefenstahl created a filmic masterpiece, *Triumph of the Will*, on what was effectively an arts grant from Hitler's Nazi government. Another racist propaganda film, *Birth of A Nation*, is also a major cinematic creation, despite its abhorrent elevation of the Klu Klux Klan to heroic stature.

S & M (SCIENCE & MORALITY): PHYSICISTS PLAYING AND ETHICS

Just as artists most often thrive in a cultural environment that doesn't interfere with their enterprise, so too do scientists. Furthermore, scientists, even more so than artists, are concerned more with their creative endeavours than with issues of morality.

.

In many ways scientists have had it much easier than artists, at least in recent times. Since so many artists now have donned the mantle of social critic, naturally enough they receive little support from those they criticize. Scientists, on the other, hand are seen as creating products that are potentially useful to the establishment. Doing modern science is extremely expensive, and scientists have become very dependent on funding from business and government to continue their work.

Of course it is technology, not pure science that is valued by the establishment. This is probably a good place to pause and examine the distinction between technology and science, for it is extremely relevant in the forthcoming examination of the way both shape artistic creation. And at this point I should confess that when I write about science collaborating in the artistic enterprise, I'm being less than precise myself; I really mean *both* the influence of science *and* of technology. This matter will be addressed in a later section.

.

Science is both a methodology and a body of knowledge. The scientist uses the methodology of science to create scientific knowledge. This is of course parallel to the artist using artistic methods to create art—a different form of knowledge. Technology refers to practical applications of scientific knowledge. (It would be nice to believe that "the good life" refers to the practical application of artistic 'knowledge'.)

.

Technology and science have a symbiotic relationship; they feed on each other. Because they are so closely bound, it is common to make the error of treating them as synonymous. Few would notice the mistake of referring to the microscope or the telescope as a major scientific advance. Galileo Galilei and Thonius Philips van Leeuwenhoek certainly did increase scientific knowledge, but their refinements of the telescope and microscope were *technological* advances—ones that made their contributions to the *sciences* of astronomy and microbiology possible. And this is a common story.

.

If one were to talk about "brain waves" to a hardheaded scientist of a hundred years ago, you might as well be talking about your invisible friends for all the attention he would pay. He might deign to remind you that science can only deal with phenomena, with observables. But then in 1912, the Russian physiologist Pravdich-Neminsky recorded an EEG, an electroencephalogram, a record of a "brain wave". Now "brain wave" is just the colloquial term for the ever-shifting pattern of voltage changes in the brain. When a neuron 'fires', the movement of sodium ions through the cell membrane results in a voltage change on the order of 100 millivolts. To detect these miniscule voltage changes through our thick skulls one needs one helluva a sensitive volt meter, and that is what an electroencephalographic recording consists of—attaching terminals from a fancy and very sensitive voltmeter to our scalps and amplifying the signals so they can be printed out on a graph. Just as with the microscope and the telescope, a technological advance opened up vast new unexplored territories for scientific investigation. And from where did this technology come? It came from the application of scientific advances in the very different domain of physics.

.

Scientific knowledge leads to technological innovation which leads to an expansion of what the scientific method can study which leads to more scientific knowledge which leads to more technological innovation which—well, you get the idea.

.

So, technology is by definition useful, and its usefulness isn't limited to the advancement of science. It's useful to industry and to governments and, of course, to us in our daily lives. There is no difficulty in recognizing the usefulness of most technological innovations. The difficulty arises in recognizing the potential usefulness of scientific discovery. Science is, by itself and in its pure form, just like art in being fundamentally useless.

.

What purpose does art serve? The answer to that question has concerned many thoughtful people, not least of all artists themselves—and, of course, the philistines as well. One is less likely to find anyone asking: "What purpose does science serve?" Why is that?

.

I have argued that both are ways of apprehending reality, ways to knowledge. Why is the knowledge gained by viewing a painting of a seascape considered useless, while the knowledge gained by knowing that water is composed of two hydrogen atoms and one oxygen atom considered useful? The former actually *feels* more relevant than the latter for most of us—as it most certainly does for the majority of students in elementary science class.

The answer? Scientific knowledge can be used to create technology. It may seem a long, long way from the abstract, and often counter-intuitive, findings of atomic theory to anything useful. Or I should say: it may have once upon a time seemed so. It doesn't anymore. No one ever built an atomic bomb from a painting.

Over and over again scientific knowledge, even in its most esoteric guises, has proved itself a potential source of powerful practical products. Few are those who now would question the maxim that "Knowledge is power", especially scientific knowledge.

But some are also maintaining "ignorance is bliss"—or at least safer than having too much powerful knowledge. Many reasonable people worry about this power and some would go so far as to try to curb it. One example is the opposition to stem cell research and genetic engineering so very much in the news. More than a few people would like to censor the publication of any part of the genetic script, even though they probably don't even know the four letters of its alphabet. Science is not alone, of course, for art has always had would-be censors to deal with. But in one sense, the stakes are higher in science, and scientists themselves are expressing doubts about the inherent and unquestionable good involved in delving deeper and deeper into the Pandora's Box the scientific method has pried wide open.

"Knowledge is power." That is empirically validated. "Knowledge is good." That seems to be open to discussion. And if one should suggest that knowledge always *results* in good, there is empirical evidence—and it does not support the hypothesis.

In 1962 the monk and poet, Thomas Merton, published an emotionally devastating poem about the bombing of Hiroshima, entitled "Original Child Bomb"*, which somehow both captures the

---

* The poem is so named because the Hiroshima bomb was named "Little Boy".

horror *and* the innocence surrounding this terrible event. I will never forget reading this poem for the first time, and I would recommend it to anyone who wants to try to understand the ethical issues raised by the whole scientific enterprise. (There are many other poems that deal with this atrocity, but I know of none that go beyond mere condemnation.)

.

It may be that advances in biology, not physics, will lead to even more potentially destructive technology than atomic weapons, but at least for now the Nuclear Bomb is the strongest evidence of the dangers inherent in scientific creativity unfettered from moral constraints. An examination of the whole enterprise, and especially the individuals that contributed to it, from Einstein down through the scientists who worked on the Manhattan Project, says much about science and ethics.

.

After the validation of Albert Einstein's General Theory of Relativity in 1919, Einstein became the most famous physicist in the world and his contributions to science considered among the most 'pure', in the sense of being counter-intuitive, obscure, abstract, and seemingly without any relationship to the 'real' world. Einstein, however, was fully aware of the real world implications of his theories, especially his earlier one that stated the simple equation that mass could be converted into energy at the incredible ratio of the square of the speed of light.* In 1939, when the Nazis were increasing their expansion, Einstein (who had escaped to The States) co-authored with another brilliant refugee scientist, Leo Szilárd, a letter to President Franklin D. Roosevelt warning, "extremely powerful bombs of a new type may thus be constructed" using nuclear fission and urging governmental funding of research in this area.

.

Roosevelt listened. He formed the "Uranium Committee" to investigate the potential of developing nuclear fission bombs. This, after a rather slow start, accelerated into the project that is now usually referred to as The Manhattan Project. In 1942, the physicist Robert Oppenheimer was put in charge of research and convened a conference on "nuclear weapon design", and promptly recruited most of the best physicists in the country—or who could be brought to the U.S.

.

---

* This is actually just a footnote in one of his important papers of 1905. It, contrary to the popular belief, is not central to Relativity Theory.

The work of developing the first the atomic bomb is obviously a technological endeavour, but the majority of scientists involved seem to have brought their 'pure scientist' attitudes along with them to this project. The principles of theoretical physics were being tested in this enterprise, and it seems the scientists were more concerned with this pure science aspect of the project than with any ethical considerations. Understandably, few seemed to doubt its moral justification as part of the war effort and the possibility of the Nazis developing and using just such a weapon against the Allies. Joseph Rotblat, eventual recipient of the Nobel Peace Prize, was the only scientist to quit the Manhattan Project on the basis of moral objections to the development of such a horrific weapon of mass destruction.*

.

Richard Feynman perhaps best exemplifies, albeit by exaggeration, the attitudes of those brilliant men who applied their minds to developing the most momentous and frightening technology mankind has ever produced. Feynman, from all accounts, including his own, seemed to take Los Alamos primarily as an opportunity to solve problems, play with ideas, and even play with the security forces—as previously described. Niels Bohr, the father of atomic theory, was also involved in the project, and the young Feynman was delighted to have the opportunity to boldly engage the great man in debates over fundamental principles of physics.

.

Now there is no doubt that the scientists of The Manhattan Project were men of principle and believed that what they were doing was justified given the war effort. Many, in fact, upon witnessing their labours bear the bitter fruit of Hiroshima, Nagasaki, and the subsequent Nuclear Arms Race renounced further development of nuclear weapons. Both Einstein and Szilárd, who in some sense might be 'credited' with starting the nuclear arms races by their letter to President Roosevelt, actively campaigned against nuclear proliferation. In 1955 Einstein, in collaboration with Bertrand Russell issued a 'Manifesto' warning that the development of nuclear weapons threatened the continued existence of the human race. Even Oppenheimer, commonly called "The Father of The Atomic Bomb" and libellously labelled "Oppenheimer The Million Killer" by Kenneth Rexroth†, spent the years after its horrific application

---

* Before leaving the project to return to his adopted home of Great Britain, he was accused of being a Soviet spy, and was subsequently barred from the United States for several years later.

† This in the poem "Thou Shalt Not Kill", where he also maligns Einstein.

lobbying for international control of atomic energy and a stop to the nuclear arms race. Having seen what he and his colleagues had given birth to, he said sadly: "Physicists have known sin and this is knowledge they cannot lose." (He was 'rewarded' for his efforts on behalf of saving the human race from nuclear annihilation by the U.S. government having his security clearance revoked.)

.

Passing moral judgement on the individuals involved in developing the atomic bomb is not relevant here, but what is relevant is that The Manhattan Project is a dramatic example of the fundamental *amorality* of creativity in the sciences—at least in 'the heat of battle'.

.

It is scary to think that those scientists at Los Alamos were *playing*, were 'experimenting'—just like little Johnny when he puts salt in mother's planter to see what happens to the plants. Consider the following. Before they pushed the button at Trinity Site to explode the first atomic bomb, scientists were placing bets on what would happen. These ranged from a complete dud through 18 kilotons of TNT (the hopeful prediction) to destruction of the state of New Mexico—and on even to ignition of the atmosphere through atomic fusion with nitrogen, this last resulting in the incineration of the entire planet!*

---

* This last possibility was suggested by Teller, one of the few scientists who did not eventually lobby against nuclear proliferation, and who is said to be the model for Dr. Strangelove in the film of that title. Allegedly almost all of the other scientists considered this outcome not just unlikely, but really impossible—based on careful calculations before the button was pushed. Still.

DECADENT ARTISTS AND MAD SCIENTISTS

The reason the creative act in both art and science is fundamentally amoral is that it is a drive, like that for food or water or sex. Drives, when sufficiently intense, override any moral reservations in us mere mortals—at least for those of us who are not saintly or Socrates. This is a sad fact of life: Needs, if sufficiently strong, trump ethics. I do not steal food whenever I become hungry, even if I have no money to purchase it. But if I was truly starving, and theft was the only way to survive, I'm sure I'd resort to shoplifting at the grocery store. The point here is not the philosophical question of under what conditions a normally unethical act is justified or mitigated. My point is that the creative drive can be as strong as any biological drive and like those drives is only restrained by higher moral considerations, constraints which are *external* to it. And when the creative drive is sufficiently strong, those considerations usually are conveniently forgotten and fall by the wayside—except in those of truly exceptional character.

.

This is not to imply that all creative people are amoral. On the contrary, I think the historical evidence suggests that most of the eminently creative display an exceptional degree of concern with morality. Our most brilliant writers and scientists quite consistently commit to promoting human rights and dignity and concern with the fate of our planet. They are the first to oppose tyranny, assaults on human freedom and dignity, and irresponsibility in our guardianship of the earth. They often risk their own freedom and dignity and well being in doing so, often being persecuted or driven into exile for their efforts. Artists and scientists consistently strive to climb to high moral ground, even if sometimes that ground shifts or their moral compasses are flawed. (And, very often, they may have difficulty applying their moral beliefs to their behaviour in their private lives.)

.

The creative probably are more, not less, moral than the majority, but they are also gifted and cursed with another drive that, like all drives, can undermine morality. And there are profound differences in the way this drive operates in the arts and in the sciences. The negative stereotypes associated with each epitomize this: the decadent artist and the mad scientist.

.

The Decadent Artist

.

The tabloids thrive on pointing to the clay feet of the rich and famous. Movie stars and other banal celebrities, whom the readers of the tabloids obviously envy, are the targets of virtually every piece published. For those with somewhat more elevated brows, the biographies of artists can serve the same function. The audience for both is titillated by the explicit descriptions of the personal immorality of those they envy and often, paradoxically, deeply admire.

.

The most egregious 'highbrow' example of this I've come across is *Intellectuals* by Paul Johnson, a book that digs up the dirt on everyone from Jean-Jacque Rousseau and Leo Tolstoy through to Bertrand Russell and Jean Paul Sartre, wets it down with soggy prose, and then slings this mud at them with a vengeance. (This is one book that could justifiably be called "dirty".) Johnson, a reactionary historian and journalist, tries to discredit any socially concerned writer left of Attila the Hun by unabashed argument *ad hominem*. It really need not be said that one can't really discredit a humanistic philosophy by pointing out that a proponent of this philosophy cheated on his wife or never contributed to charities. (It is pleasing to note that Paul Johnson was eventually hoisted by his own petard. If his brand of conservatism legitimately could be repudiated by his own life, it is very dead meat indeed. This confessed admirer of General Franco and dining buddy with General Pinochet, this defender of Richard Nixon and evangelical creationism, has had his own personal life come under scrutiny. Suffice it to say it ain't pretty, albeit sometimes it is funny.*)

.

We all, at very least after passing through puberty, have some skeletons in our closets. And this effectively means that we are *all* hypocrites whenever we promote moral behaviour. Virtually no one has always practised (or still always practises) what he preaches! As the ease of 'data collection' on each of us increases, so too does our vulnerability to having our heartfelt convictions and even our contributions to civilization discredited by the exposure of our all-too-human foibles. Those few who would enter politics out of real concern for the commonweal hesitate far more now that their private lives are so subject to public exposure. The price for fame or a life in the public eye has gone way up: Fewer and fewer are willing to pay it. But artists still do.

.

---

* A number of liberals call him by the code name "Spanky".

I have talked about drives. Some psychological jargon here might actually clarify the discussion. According to traditional motivation theory, all organisms strive toward homeostasis. This word means balance. When things get out of balance, the organism feels a *need*. The urge to fulfill this need is called a *drive*. When the drive succeeds in fulfilling the need, homeostasis is temporarily restored. Most of this is handled by a structure in the brain called the hypothalamus, which also is known as the 'pleasure centre', for it recognizes biological needs, initiates drives, and rewards the organism with a feeling of pleasure when these drives result in the needs being satisfied. It tells us we're hungry and need to eat, drives us to seek food, and rewards us with pleasure when we find food and eat it.

Even the most sophisticated brain-scan technology has failed to locate the part of the brain that detects a need to create, drives the creative person to do so, and rewards the completion of the creative process. Nevertheless, the analogy is still appropriate. The creative are definitely *driven* to create by a need to do so—and find it rewarding, pleasurable, to do so. But the important thing to note here is that *finding* food is not what is rewarding: it is the *consumption* of the food and the resulting homeostasis that affects the pleasure. Food itself is merely prerequisite to the rewarding act of consumption. With artistic creativity, there are numerous prerequisites to the rewarding act of creation. Among these are emotional and intellectual arousal, as well as diverse experiences. Finding these can involve trampling on convention and even on the lives of others.

To the artist, all experience is grist for the creative mill. Experience is a merely a means to an end. Contrary to the idealized conception of the artist savouring the experiential world, the artist is really just taking notes, making sketches, planning some work. The artist does not bask in the moment, for he can't help but see it as a means to an end. Like the tourist walking around seeing his holiday through the viewfinder of his video camera, the artist is often too busy recording to experience and appreciate. The innocent wide-eyed child stops to admire the flower. The artist stops to sketch it, capture it. The sensitive, caring daughter weeps at her mother's funeral, but if she is a writer, she is also taking mental notes on her own feelings for use in some future book.

The creative artist needs experience, just as the hungry person needs food, as prerequisite to fulfilling a need. For the artist the drive to find experiences, especially experiences that are intellectually and emotionally arousing, can be extremely intense—and disruptive to the status quo, the artist's friends and family, and the artist's own life.

There is a 'personality disorder' in the DSM-IV taxonomy called "Histrionic Personality Disorder" No matter whether this is or is not a legitimate 'mental illness', it certainly does describe a certain type of person. 'Histrionic' technically simply refers to having to do with the theatre, but the adjective implies melodramatic or exaggerated expressiveness of emotion—such as required of stage actors to project emotional states to the audience. The 'histrionically disordered' person attempts to turn every quotidian fortune or misfortune into high drama. "I missed the bus. 'I fall upon the thorns of life. I bleed!'"* This suggests an abnormal need for intensity in one's emotional life. The same need seems to be present in artists and usually is not so easily satisfied by subjectively ramping up the amperage of everyday experiences.

Gustave Flaubert advised other writers to "live like the bourgeoisie", so as to conserve their emotional energy for their creative work, but few writers seem to heed his advice.† It is intense emotion that fuels their fire, and they often seem to deliberately upset the bourgeois applecart of their lives to produce it. The image of the writer only writing when inspired by great and deep emotional experiences is of course naïve: writers, even poets, often go off to work quite grudgingly and feeling no more intense emotions sitting down at their desks than do the average working-stiffs dragging their asses to the factory. But there *is* a grain of truth in the old and conventional idea of 'inspiration' resulting from profound emotional experiences.

Of particular interest to the readers of the tabloids are the sex lives of the rich and famous. This isn't surprising since matters of the heart and groin, second only to illness and death, give rise to the most intense emotional experiences of our lives. For most of us the chosen path, the safe path, is the vicarious one, be it National Enquirer exposés or novels or biographies. But the artist cannot rely entirely on secondary sources. Henry Miller could not have written *The Tropic of Cancer* solely on the basis of reading raunchy novels and travel guides to the seamy underside of Paris.

---

* Apologies to Percy Bysshe!

† Interestingly, this great novelist truly hated the bourgeoisie, and his relatively uneventful life (by objective standards) apparently so stressed him emotionally that his health deteriorated at an early age; reputedly he looked and acted like an old man before his 50th birthday. He died at age 58.

What I am saying does not apply only to writers, for composers and visual artists, artists in all genres, need experience to stoke their fires. Writers may be the ones who most directly use their experiences as material to shape their work, but Picasso's oeuvre surely would be less notable had he "lived like the bourgeoisie", and Berlioz's revolutionary *Symphonie Fantastique* would never have been composed had he not hung out with a 'fast crowd' at the theatre and 'lost it' over a second-rate actress.

Seek experience. Experience equals opportunity. One synonym for opportunity is temptation.

Celebrities have opportunity on the basis of their celebrity. Artists have opportunity on the basis of actively seeking experience. So it is not surprising that both often give in to temptation and so supply great, juicy material for biographers—and muckrakers.

Hence the stereotyping of the artist as decadent, just as those associated with the theatre, as decadent: this is a stereotype, like most, not entirely without the proverbial grain of truth.

I should conclude by admitting that in this analysis I occasionally may be guilty of the correlational-thinking error of confusing cause with effect. No doubt in many cases the causes of the intense experiences that inspire the artist are not of his own making; they may be mere accidents of fate, chance events that initially light the fire of creativity within him. But whether self-inflicted or caused by The Fates, the fact remains that highly charged personal experience is the fuel for that fire. This is why most artists' biographies (once the long, solitary hours at desk or canvas are edited out) seem so chaotic, histrionic, complicated, and rich—albeit often tragic. And appear to the envious bourgeoisie so deliciously decadent and immoral.

The Mad Scientist

The New Scientist (24, September, 2005) published a survey of one thousand 'horror films' shown in the decades from 1930 to 1980 that found mad scientists to be the villains of 30% of these flicks, with scientists being the heroes in only 11%. Popular culture mirrors and caters to popular prejudices and fears. Most fears and prejudices are based on ignorance. So given the appalling ignorance of science in the general populace, it is not surprising that many people fear scientists, while simultaneously and unashamedly revelling in the technological spin-offs of scientific research.

.

Some would say it all started with Mary Shelley's novel *Frankenstein*. Of course it actually started with the Romantic Movement in the arts, which one could argue first opened the chasm between art and science, and this book—this wonderful and brilliant book—was 'only' the first extended narrative expression of this fear of the mysterious and potentially destructive power of science. Mary Shelley was married to Percy Bysshe and part of the whole extended literary 'family' of Romantics that included Keats and Byron. They all feared science, for they felt it could murder the joyous mystery of life, dissect the rainbow, leaving nothing but jagged shards on the lab counter.

.

Another fear associated with the scientist, if not one only associated with the Romantic Movement, is that the scientist was playing God— and without God's wisdom. Enter The Church. The two very different and very strange bedfellows intent on demonizing scientists were the artists associated with the Romantic Movement and religious folk. Both felt hubristic barbarians in lab coats were invading their territory. The former felt this because they feared science would sully the mystery and magic of art. The latter felt this because they feared it would sully the mystery and magic of religion.

.

I suppose it has to be admitted that their fears were in some ways justified, although certainly not in the ways they envisioned. As I argue as the main thesis of this book, science adds more to art than it subtracts, but it does undermine old unsubstantiated beliefs. And science indeed very successfully undermines traditional religious dogma—but that is a commendable accomplishment, in my view.

.

Science does not diminish the mystery of the world so much as it enriches and adorns it, but then, as now, this was not fully recognized. Even as we harvest the fruits of applied science in the form of technology or a broader canvas for artists, we still remain suspicious of it. Our fruits and vegetables come to our tables more, not less, fresh and nutritious than in the past (unless we grew them ourselves), yet fear of 'genetically modified' food is as rampant as once was fear of witches. Paranoia about mysterious radiation from cell phones causing cancer is almost as common as the use of cell phones.

.

Yet we do not trust the magic and especially the magicians—for scientists are the magicians of our time, performing feats that are incomprehensible to the average person. Our fear and suspicion is deepened by the knowledge that they have so drastically changed our lives. This distrust is partially based on ignorance of the methods of science and those who use these methods. But that is not to imply that one *should* always trust these modern magicians, only that one should know enough to know when to trust and when not to trust. Most scientists are smart but not necessarily wise. And some scientists now too readily accept playing the role formerly assigned to some imagined deity.

.

But to return to evaluation of the validity of the stereotype of the mad scientist, the first thing that needs correction is the image of the lone scientist, with wild hair and wild eyes, toiling away in a subterranean laboratory with Van de Graff generators sparking in the background. For one thing (and for better or for worse) science has become a collective endeavour. The individual scientist working alone is a thing of the past. Most research now requires expensive equipment and such equipment is funded by grants to institutions (academic and corporate), not to independent scientists. Most scientists work in collaboration and with institutional support. Einstein may be the last great scientist to truly work alone and outside of some establishment.

.

Other more trivial aspects of the stereotype, such as the European accent and the absent-mindedness and the strange habits and social ineptitude have—or once had—a grain of truth to them. Yet the crux of the stereotype is the mad scientist's desire to 'play god', without the moral wisdom people usually attribute to their Gods.* The previous section addressed the question of the morality or amorality of scientists as they ply their trade.

.

It is useful to contrast the artist and the scientist in terms of morality. Both need experience to feed their creative fires. Both seem willing to take risks, with their own lives and others, to gain that experience. But for the artist, the risk is primarily to himself and those close to him, for it is personal experience he seeks. For the scientist, the risk

---

* Although only God (?) knows why, for most gods of all religions behave abominably by any reasonable standard of ethics, including the ones they pass down as commandments. "Thou shalt not kill!" Good idea, God, but then how to justify the largest genocide in the human history—the Flood you allegedly sent to wipe out the human race minus a drunk and his family?

can be much greater and far reaching, for it involves putting at risk much more, even potentially the future of our species or the planet. The test of the atomic bomb, so aptly named Trinity, was an experience devoutly desired by the scientists at Los Alamos but involved a risk far greater than that of an artist abandoning bourgeois existence and family to live in Tahiti and paint pictures.

.

All creative people 'play God' by the mere act of creation. The novelist invents people, defining their character as no parent ever could a child, subjects them to all manner of trials and tribulations, and often even kills them off, all this without even the tiniest pang of guilt. But the scientist is playing god on a larger stage: the real world. This may not be madness, unless creativity itself is madness, but it is hubris that, were there really omnipotent gods, would surely be disastrous for us mere mortals.

.

One can only say: so far, so good, and may sanity prevail.

CASE STUDIES: VICTOR FRANKENSTEIN, HENRY MILLER

Huh? Dr. Frankenstein is a fictional character. Hey, that doesn't mean he isn't worth analyzing. He has a biography. And, besides, the Henry Miller of the autobiographical novels is fictional too. At least I'm being consistent.

.

"It was a dark and stormy night…" This is the cliché beginning of a gothic tale, and now is even the name of a competition for creating the worst first line of a popular novel. But apparently it really was a dark and stormy night that Lord Byron suggested to his companions a little literary contest to see who could write the scariest story. This was in 1816 and the volcanic eruption of Tambora the previous year had darkened the skies over Europe, so that even midsummer was cold and dreary. Lord Byron was playing host to Percy Shelley and his fiancée Mary Wollstonecraft Godwin at his villa by Lake Geneva in Switzerland. (Byron's personal physician, John Polidori, was also there.) They'd been reading an anthology of German ghost stories, apropos the dreary weather. Polidori wrote *The Vampyre*, one of the first vampire tales, but the clear winner of the contest was Mary Godwin (soon to be known as Mary Shelley) for her *Frankenstein: The Modern Prometheus*.

.

Most people know the general plot of *Frankenstein*. Although the original version, the edited version and the various movie versions are in many ways different, the central idea of the story and the character of Doktor Victor Frankenstein is similar enough in all variations to permit analysis.

.

Dr. Victor Frankenstein is the embodiment of what the Romantics most feared about the scientist. His scientific understanding had equipped him with the power of God: He'd figured out how to create sentient life. By piecing together cadaverous body parts he shaped a human-like figure, and then imbued this creature with the *élan vital*. This is what God did when he created Man. The doctor is laying claim to being a creator in the ultimate sense that religious people refer to God as *The* Creator. Of course this what many fear is really going to happen now, if the biological sciences are not held in check. Or even what could happen if artificial intelligence research eventually resulted in sentient beings.

.

Dr. Frankenstein is the archetypal mad scientist. In his incredible hubris, he does not consider the moral implications of what he is doing. The amorality of scientists driven by their creative urges is epitomized in Frankenstein. Still, he is not an immoral man. He is not a bad person. In some sense, his urge to create life is similar to the universal procreative urge to have children, children who we can shape to fit our own conception and continuation of ourselves. However, he uses his knowledge of natural science to create what is unnatural, and this is what frightens us still. Fusion in the core of our sun is natural, but when we create fusion ourselves it takes the form of something as unnatural and horrific as the hydrogen bomb.

.

Dr. Frankenstein succeeds in creating this living creature, as innocent as a natural newborn babe, but somehow unnaturally hideous to his father. As his son comes to life, he is struck with horror.

.

"How can I describe my emotions at this catastrophe, or how delineate the wretch whom with such infinite pains and care I had endeavoured to form? His limbs were in proportion, and I had selected his features as beautiful. Beautiful!--Great God! His yellow skin scarcely covered the work of muscles and arteries beneath; his hair was of a lustrous black, and flowing; his teeth of a pearly whiteness; but these luxuriances only formed a more horrid contrast with his watery eyes, that seemed almost of the same colour as the dun white sockets in which they were set, his shrivelled complexion and straight black lips."

.

And then eventually like Oppenheimer, and most of the others who worked on the Manhattan Project, Frankenstein repudiates his creation.

.

"I beheld the wretch -- the miserable monster whom I had created. He held up the curtain of the bed; and his eyes, if eyes they may be called, were fixed on me. His jaws opened, and he muttered some inarticulate sounds, while a grin wrinkled his cheeks. He might have spoken, but I did not hear; one hand was stretched out, seemingly to detain me, but I escaped, and rushed down stairs. I took refuge in the courtyard belonging to the house which I inhabited; where I remained during the rest of the night, walking up and down in the greatest agitation, listening attentively, catching and fearing each sound as if it were to announce the approach of the demoniacal corpse to which I had so miserably given life."

.

The effects of Frankenstein's rejection of his artificial offspring are the reason he comes to be called "monster". This creature comes into the world with the same needs for love and affection and education that any child does, but his rejection is what scribbles on his blank slate ugly emotions such as jealousy and hate, for initially Frankenstein's 'monster' is no monster.

.

Victor Frankenstein is guiltier for repudiating his creation than for the act of creation. The real madness in scientists is not the mad drive to create, but rather the madness of not taking responsibility for what they create.

.

And what about the amorality of the artist, intent on creating at any cost, but not willing to take responsibility for the damage he does while doing so? Henry Miller is not Frankenstein's monster, but he is in many ways the artistic analog of Victor Frankenstein. And, it is worth noting that his amorality is part of the same Romantic tradition that was so fearful of the scientist's amorality: Byron, et al were throwing stones out the windows of their glass houses.

.

Henry Valentine Miller was born on Boxing Day in 1891. He grew up—if one can apply that phrase to him—in the tenements of working-class Brooklyn. The rough and ready streets of this part of New York at the turn of the century were a goldmine of material for a writer, but writers didn't *live* there. For a young man from Miller's neighbourhood to even conceive of 'being a writer' was considered the height of lunacy and/or immaturity—roughly equivalent to stating as your vocational plans as becoming President of the United States or at very least Harvard University. Learning a trade, perhaps becoming a tailor, like Miller's father—now that was more realistic.

.

This is not to imply that no intellectual life could be found or created in this environment, for more great writers come from so-called 'humble' backgrounds than from wealthy suburbs. It is to say that Miller did not have the educational advantages and opportunities so casually afforded, and so often unappreciated, that even the working class youth of today have. So he created his own intellectual subculture, even forming a literary club, called the "Xerxes Society", and hanging around the more bohemian and disreputable types, such as theatre people. He was very much an autodidact and a voracious reader. He briefly* attended City College of New York, but, like most

---

* Two months!

writers, found the academic approach to literature unbearable. He says it was studying Spenser's *Faerie Queen* that convinced him to pack it in and follow his own idiosyncratic interests and tastes in reading material. While profoundly passionate about literature, his eclecticism is disconcerting to most traditional critics. He writes as passionately and reverentially about H. Rider Haggard (author of the charming pot-boiler *King Solomon's Mines*) as about Friedrich Nietzsche.*

In the thirties Miller moved to Paris, just as the Lost Generation literati of the Jazz Age Twenties moved out. Unlike the expats that had moved on, Miller did not travel in sophisticated literary circles, take little forays to Spanish bullfights or attend literary soirees hosted by the likes of Gertrude Stein. Rather he lived in dire poverty and hung out with a very motley crew of misfits and outcasts. He eventually made contact with some other outré writers, including Anais Nin. It was her financial support that led to the publication, in 1934, of his first book, *Tropic Of Cancer*. Miller was already into his forties.

If you mention the name of Henry Miller to folks who have not read him, but recognize his name, it is this book they will have heard about. It is this book that eventually made him famous, although—or because—it was banned for three decades in the United States! The 1964 Supreme Court ruling that deigned to allow publication on the basis of "artistic merit" is one of the landmark freedom-of-expression decisions, along with the earlier 1933 decision regarding James Joyce's much tamer *Ulysses*.

Tropic of Cancer is infamous and important. It is not his best book, by any means.† It is not pornographic, if that ambiguous word implies being erotic. It is, however, a perfect example of the amorality of art and artists and one of the most painfully honest books ever written. This book, as Miller writes "…is not a book. This is a libel, slander, defamation of character. This is not a book, in the ordinary sense of the word. No, this is a prolonged insult, a gob of spit in the face of Art, a kick in the pants to God, Man, Destiny, Time, Love, Beauty."

---

* One of his books, a book that guided my own reading more than any traditional curriculum, is called *Books In My Life*. Like Rexroth's *Classics Revisited* I think it should be required reading for anyone who presumes to teach a course in 'literature'.
† I personally rate his *Colossus of Maroussi* near the top, and this charming memoir about Greece has nary a naughty word or sexual encounter.

Not surprisingly, those he libels, slanders and defames (although almost always in the abstract) loathe the book. Who wants to have some foul-mouthed, indigent, promiscuous, uncredentialed smart-ass spitting in their face? Academics did not deign to consider Miller a 'real' writer. Feminists demonized him. Puritans wanted his mouth washed out with soap and his books not just banned, but burned. (But he has had his admirers. One can judge a person by the quality of their enemies, but also by the nature of their admirers. Kenneth Rexroth, Norman Mailer, and Karl Shapiro have all written extensively about Miller's importance.)

.

The reason Henry Miller is considered such an important writer is well described by Rexroth in his introduction to Miller's *Nights of Love and Laughter*.

.

> *"His absolute freedom from the Christian or Jewish anguish of conscience, the sense of guilt, implication, and compromise, makes Miller humane, maybe even humanistic, but it effectively keeps him from being humanitarian. He might cry over a pet dog who had been run over, or even punch the guilty driver in the nose. He might have assassinated Hitler if he had had the chance. He would never join the Society for the Prevention of Cruelty to Animals or the Friends Service Committee. He is not involved in the guilt, and so in no way is he involved in the penance. This comes out in everything he writes, and it offends lots of people. Others may go to bullfights and write novels preaching the brotherhood of man. Miller just doesn't go to the bullfight in the first place. So, although he often raves, he never preaches."*

.

Miller was profoundly amoral in his embracing of life[*]. He accepted life as it is—and not surprisingly makes frequent reference to the Buddhism. He looks at the Emperor in his new non-existent clothes with the amoral innocence of a child and gleefully reports on (as Rexroth puts it) "the pimples on his behind, and the warts on his private parts, and the dirt between his toes". But many other writers have done that. What distinguishes Miller is that he looks in the mirror and reports just as gleefully on his own pimples and warts.

.

For Miller, his own—sometimes sordid, sometimes bathetic—life was the material from which he shaped his books. These books, labelled 'autobiographical novels', are not objective documentary

---

[*] Miller has been compared to poet Walt Whitman in this regard. Not only did both totally reject the conventions of their genres, but they both totally rejected the conventions of their culture.

records. They are full of exaggeration, surreal flights of fancy, sophomoric and garrulous philosophizing, as well as lots of warts and pimples. But unlike virtually all autobiographies, they do not attempt to make the author look good.

Henry Miller was not interested in looking good. He was not interested in being—or pretending to be—of high moral character. He *was* interested in shaping books from his own chaotic life, without any cloaks, pretences, or self-justifications. In this way, he epitomized the amorality of the artist in the throes of his creative urges.

I should conclude by emphasizing that in calling him amoral, I do not mean he was immoral. (He is no more immoral than was Doktor Frankenstein.) The silly comparisons of Miller's work to that of the Marquis de Sade are just that—silly. All the real biographical evidence suggests that Henry was a very nice man and would have made a good friend.* If amoral folk like him were running things, rather than the putative moralists now in control, we'd all probably be living in a happier, safer world.

Henry Miller lived to a ripe old age. In 1940, he left Paris and returned to the United States to a cabin in Big Sur, California. Here he continued to write (and paint watercolours) and live on the edge of society and on the kindness of strangers—and friends. He eventually moved to Pacific Palisades, where he died, at age 89, on June 11, 1980.

---

* Although you, if a man, might not want to trust him with your wife. And, if a woman, expect to be treated like a man. (It is interesting, that for all the alleged sexism in his writing, there are few documents from the women in his life that portray him in negative light.)

MADNESS AND GENIUS: ALLIES OR ENEMIES?

"The creative person is both more primitive and more cultivated, more destructive, a lot madder and a lot saner, than the average person."
—Frank Barron (article in *Think*)

.

"There is no great genius without some touch of madness."
—Seneca (*Epistles*)

Whether or not madness is linked to genius, or what we now call 'mental illness' linked to what we now call 'creativity', is an old question, and, as in the nature versus nurture debate, no side is ever going to claim a unanimous decision by objective judges. But historically the odds makers have favoured nature as the stronger force in the one debate, and madness really being connected with genius in the other.* Current scientific evidence tends to support these millennia-old opinions.

.

The primary difficulty in dealing with the issue of madness and genius is that both terms are difficult to define. And both terms have many layers of often conflicting connotations laid upon them—with those contemporary ones of psychology and popular usage being the most confounding. I will attempt to operationalize these definitions—or at least be clear as to what I mean in each case I use these terms.

---

* In Ancient Greece, and other ancient cultures, it was commonly assumed that all genius was tinged with madness—as in the epigram by Seneca, which is supposedly based on Aristotle saying there was never a genius without a tincture of madness.

CASTLES IN THE SKY: NEUROSIS AND PSYCHOSIS

The historian Edward Gibbon said, "There is more pleasure to building castles in the air than on the ground." But very few living in those airy castles seem to take much pleasure in their residence—or accomplish anything. It should be obvious that if neurotics build castles in the sky and psychotics live in them*, the former are the creators, the latter just unhappy tenants. The most convincing refutation of the idea that creative people are mad *in the sense of being psychotic* is a trip through the chronic-care wards of any mental institution. The unfortunate residents one would encounter there are incapable of taking care of even their most basic, animal needs—never mind having the presence of mind to compose a symphony or a poem or a work of visual art. The truly mad are as incapacitated as are the truly retarded. The insane asylum is no artist studio or study, no scientist's laboratory.

Nevertheless, Aristotle's suggestion, echoed by Seneca, that a "tincture" of madness seems to be a necessary ingredient in creative genius does have some support, as mentioned in the discussion of personality in an earlier chapter. The problem is, yet again, one of definition. What is 'madness'?

The taxonomy of mental illness is extremely controversial and has undergone many, sometimes drastic, changes over the last hundred years. And it continues to change. As mentioned in the previous section, the official 'Taxonomical Bible' of mental disorders is the *Diagnostic and Statistical Manual of Mental Disorders*, now in its fourth full revision. The number of changes and the seemingly arbitrary structuring of this standard reference work for clinical psychologists and psychiatrists make many scientists in other disciplines sceptical of its external validity. While it is true that plant and animal taxonomy has also undergone substantial revision since Carl Linnaeus published the first edition of his classification of living things, *Systema Naturae*, in 1735, his basic outline has survived the test of time. And, more importantly, the categorizing criteria are clear and objective. This is not true—or at least far less true—of the taxonomy of mental illness.

Nevertheless, one apparently clear division, perhaps analogous to plant versus animal, is that between neuroticism and psychosis. A

---

* And the saying sometimes continues: "And psychiatrists collect the rent."

48

person suffering from the former is functional, capable of caring for themselves and being productive. Neurotics may be miserable, inefficient, and prone to making a mess of their lives, but they remain capable of accomplishment. And some neurotic behaviours may in fact augment their productivity: notable examples being a tendency to obsessive-compulsive behaviour (which leads to perfectionism) and hypersensitivity or neurasthenicism (which leads to acuity of insight).

.

It would be nice and simple to conclude that the creative are not, cannot be, psychotic, although they may indeed tend more toward behaviours that are conventionally classified as neurotic than does the average person; and that this tendency is more prevalent in artists than scientists, although present in both. This is, unfortunately, too pat.

.

Psychosis, or full-blown insanity, is not necessarily chronic and continuous. Sometimes it is, but sometimes it is not. One can go mad, become totally incapacitated, useless to oneself and utterly incapable of producing anything—and then recover and even use the experience of madness as material for creative construction. Then, in many cases, go stark raving mad again. To say that "life has its ups and downs" for such people is a pathetic and comical understatement.

WHAT GOES UP MUST COME DOWN: MANIC-DEPRESSIVES

In 1993 Kay Redfield Jamison published a book entitled *Touched With Fire* which argued quite persuasively that there is a strong correlation between bipolar disorder (previously called manic-depressive disorder) and creativity. Bipolar disorder is considered a psychosis, but unlike schizophrenia, it is not manifested continuously. Before the advent of symptom-attenuating drugs such as chlorpromazine, anyone struck with schizophrenia was effectively doomed to a lifetime of hallucinations, helplessness and hospitalization. Often psychiatrists doing highly speculative *ex post facto* diagnoses of artists whose biographies strongly suggested mental imbalance (such as Van Gogh, Lord Byron or William Blake) would label these artists schizophrenic. How they come to this conclusion is a mystery, since schizophrenia has for a long time been known to be chronically and continuously debilitating. However, as Jamison points out, this is not true of bipolar psychosis. The manic-depressive has long periods of relatively normal functioning buffering the periods of mania and depression, and, furthermore, is quite capable, sometimes exceptionally capable, of productive work during the transitional stages when the mania or depression has not yet become extreme.

Jamison is a psychiatrist and a manic-depressive. When she speaks from personal experience*, one sits up and listens with attention. When she speaks as a psychiatrist, one has to take into consideration her training and the biases inherent in it.† But unquestionably she is a sensitive and intelligent soul, "touched with fire" herself, and her book is rich with poetry—and, incidentally, far more understanding of literature than the writing of most literary critics. One has to take her and her thesis seriously.

She presents substantial evidence that the tendency to bipolar disorder is heritable. Furthermore she reviews the current diagnostic criteria for the disease and argues convincingly from biographical records that many of the 'mad geniuses' of history had the classic symptomology. The problems begin in late adolescence or early adulthood. Periods of relative stability and normal functioning are

---

* She never does this directly, although it is often obvious she speaks from personal experience. She later wrote a book 'confessing' that she was manic-depressive herself and describing her life before she started lithium treatment.
† I'm sure she takes the DSM-IV far more seriously than those outside of the field of psychiatry.

interrupted by cyclical mood fluctuations that could be quite extreme—so extreme in many cases as to require temporary hospitalization. Suicide rates are very high for those suffering from this disorder, for life is during the depression part of the cycle perceived as nothing more than unrelieved and unrelievable anguish. (The suicide rate for eminently creative individuals is indeed far about the average.) Alcoholism or other drug dependencies are common, for the victim often turns to self-medication to attenuate his or her extreme moods.

.

It is somewhat unfair to try to summarize her arguments and conclusions, for they are sometimes subtle and usually cautiously qualified. Nevertheless, Jamison believes, and offers reasonable evidence, that bipolar disorder is a genetically determined physical disease affecting brain function, although environmental factors do play a part, especially in its severity.

.

She believes the disorder is far more common in artists than in the general population. (And most prevalent among writers, especially poets.) She believes that the emotional extremes experienced by people both gifted with creative talent and cursed with the disease are at least partially responsible for these people's productivity and the profundity of their work. When neither the mania nor the depression is debilitating, the artist's aptitudes are enhanced. It should, however, be made clear that she does not maintain that all creative people are bipolar, or that the disease is any way prerequisite to genius—only that it is exceptionally common among artists.*

.

It is interesting that Jamison has commented elsewhere about whether she, given the choice now, would actually choose to have the disease. Although she in the end says she would not, she does indicate that she would feel a profound loss if the elation she felt during some of her manic phases were to be erased from her memory. And the fact that she could even conceive of such a question strongly suggests that manic-depression is not *entirely* a curse.

.

Jamison's book deals with artists and, in her examples, more often with writers than with practitioners of the other arts. What about those in other creative fields?

.

---

* Nor, of course, does she imply that people with bipolar disorder are generally more creative!

A different and very extensive study by Arnold Ludwig reviewed the biographical information of a large sample of eminent and accomplished people* in many different fields of endeavour. This sample included not just those in arts but businessmen, public officials and scientists as well. What is of interest here is that he found two to three times the rate of psychosis, suicide attempts, mood disorders and substance abuse among the artists than among the scientists. Nevertheless the rates for scientists were still well above the general population norm.†

What can one conclude from studies such as those of Jamison and Ludwig, as well as the data on personality testing, specifically 'neuroticism', reviewed in an earlier chapter on the creative personality? Well, the evidence is admittedly anecdotal and quasi-scientific, but sometimes it is reasonable to convict on circumstantial evidence. There seems little doubt that the exceptionally creative are also more likely to be exceptionally—what is a good word? Unstable? Dysfunctional? Nuts? This is true for artists even more so than scientists. But before examining the implications of this conclusion, I want to examine more closely a 'neurotic' trait of particular relevance to creativity.

---

* He used people who had been reviewed in the *New York Times Book Review* section between the years 1960 to 1990.
† Ludwig, A.M. Creative achievement and psychopathology: Comparisons among professions. American Journal of Psychotherapy, 46 (1992) 330-356.

OUT, OUT DAMN SPOT: OBSESSIVE-COMPULSIVES

I have argued in the discussion of the personality trait of conscientiousness that some degree of obsessiveness and compulsiveness is a positive value, especially in the creation of a complex thing such as a scientific theory or a novel. If Homer nodded too often, we wouldn't consider him the great writer we do.

.

Jamison remarks early in her book: "Indeed, whether out of a need to impose order upon a chaotic internal universe, or for other as yet unexplained reasons, many individuals with manic-depressive illness are inclined to be unusually obsessive and highly organized."* Well, whatever the manic-depressive's motives, the creative individual just plain *needs* to be highly organized, whether or not it comes naturally or as a further symptom of an inherited mental illness.

.

Here we run into a contradiction with the stereotype of the artist, and to a lesser extent the scientist as well. Creative people seem to exist amidst a monstrous chaos. The cluttered desk, the messy studio, is considered more typical than the neat-as-a-pin working environment of the CEO. Well, I can't offer up any empirical studies, but among the creative people of my acquaintance I've seen both incredible clutter and immaculate order, sometimes co-existing. For every Henry Miller hammering at his typewriter amidst reams of papers and books and empty wine bottles, there is a Leonard Cohen writing at an empty table in a bare white room on a sun-parched Greek Isle.†

.

Also part of the confusion may have to do with superficial appearance not reflecting reality. What looks like a mess may not be. An academic colleague of mine was remarkable for his punctuality and efficient juggling of the many responsibilities and commitments his position as Dean entailed. Furthermore, he never seemed to misplace any memo or contract or agenda, and this wasn't just because he had a good secretary. Yet his desk looked like the aftermath of an explosion in a printing factory. When he had to move to another office, he had his desk tightly wrapped in plastic so no papers slipped to the floor or shifted from its assigned position. We

---

* Jamison / p. 98.
† Personally, I try to keep my own desk immaculate, even obsessively lining up my pens in parallel next to my mouse pad, but to reach my desk on some days I literally have to jump over the piles of books and papers on the floor.

all have our own way of organizing things. What looks like mess to an outsider may just be a very efficient filing system.

I sign all my correspondence "Peace in complexity", for I believe that this is the ideal state, the unreachable, but worthy, goal of any life. Order and harmony is easy, but boring, when it only involves a few entities. To create harmony from a welter of complex and disparate elements is a damn good definition of both a great work of art and a great scientific theory—and a balanced but full life. Consider the brilliant integration of clashing, almost cacophonous, rhythms into a harmonious whole that Stravinsky accomplishes in his *Le Sacre du Printemps,* or the chaotic variety of living forms all of which Darwin was able to integrate into a single Theory of Natural Selection.

The more experiences, ideas, and emotions we let into our lives, the more we move toward chaos. But if in our art or science we can find (or make) connections and balance the light and dark components of this chaotic complexity, then we can find peace. Extreme openness and sensitivity to experience, emotional or intellectual, chosen or inherited, demands an exceptional power of organization. Sanity and survival are dependent on it. Those who go insane and those who don't survive may have had too much chaos—or too little obsessive-compulsiveness.

THE CREATIVE ACT: SYMPTOM OR SIDE EFFECT OR ACT

"Does one have to be crazy to be creative?" "Nope, but it helps!" This is meant, of course, facetiously, as in: "Does one have to be crazy to have children?"

.

However, the idea that what we have come to call 'mental illness' is often, albeit by no means always, associated with creativity is a disturbing one. Proponents of self-actualization, must have a hard time reconciling their idea of creativity as the ultimate act of existential authenticity and self-actualization with the suicides and mental breakdowns that so frequently seem to accompany reaching this 'pinnacle' of personal fulfillment. And what are the educators and self-help types who allegedly are committed to increasing people's creativity to make of this? Should De Bono include in the contracts to take his Creativity Courses a health warning? Should publicly funded schools be trying to make students more creative, given the supposed hazards associated with their being successful at it— unlikely as that may be?

.

The most obvious confound here is that of cause and effect. That writers tend to be heavy drinkers obviously does not mean that heavy drinking will make one a writer. But, on the other hand, the frequency of drug and alcohol use by artists *may* be because they do find it helpful in their work. If Jamison is right that a gene or set of genes is responsible for a physical disease that affects the emotions (i.e., bipolar disorder), and that this disease (in people gifted with talent and intelligence) augments their creativity, then would it not be civilization's loss if the disease were eradicated?* Sometimes bad things yield good things. But that is not a justification for doing bad things.

.

Putting aside the complex ethical and philosophical issues and all the psychological theories, here are the commonsensical conclusions. Creativity is associated with, and enhanced by, certain personal characteristics and intense and varied experiences. There is reasonable consensus as to what these characteristics and experiences are. The characteristics are: high levels of motivation and energy and conscientiousness; substantial intelligence and an ability to find or invent connections between disparate elements; and independence of

---

* Jamison deals with this difficult question in the final chapter of her book.

thought. The requisite experiences vary in type: they can be emotional or intellectual or actual. Regarding both characteristics and experiences, some are the result of nature, some nurture, and some a person's will.

.

For example, a writer has need of a variety of intense emotional experiences to fuel the fire to forge in the smithy of his soul a glowing work of literature. These experiences may be the result of happenstance; they may be recognized as essential fuel and wilfully sought out; or they may be the result of an inherited neurological disorder such as manic-depressive illness. No matter: these experiences are essential—where they are obtained is almost irrelevant.

.

Or consider a scientist such as Darwin. Where did he get his determination to embark on a long ocean journey when he was so prone to seasickness? From whence came his exceptional ability to find relationships where others only saw inexplicable diversity? Was it his upbringing or his genes? And would he ever have formulated his theory had not Fate offered up the opportunity to sail with The Beagle? Or consider Galileo and the suffering his independence of thought caused him. These scientists weren't mad by current conventional standards, but I'm sure many of their contemporaries thought them so.

.

If sometimes madness is the price paid for creativity, the creator may or may not have signed the bill of sale. No matter, a deal is a deal.

THE CREATIVE ACT: THERAPY OR RETREAT OR CURSE?

Dryden claimed that "Great wits are sure to madness near allied, / And thin partitions do their bounds divide."[*] If this be so, and so it seems to be, is it the practice of their art or science that keeps great wits from falling through to the other side?

.

Since artists, even more than scientists, seem prone to serious mental illness, they will be the primary focus in this discussion of the therapeutic role creating can play in their lives.

.

Probably because so many artists have publicly stated (often eloquently) that it is their art that keeps them sane, the idea of art as therapy has come to be unquestioningly accepted as reasonable. The poet Antonin Artaud expressed the feelings of a number of his colleagues (albeit with hyperbolic exaggeration) when he claimed that "No one has ever written, painted, sculpted, modelled, built, or invented except literally to get out of hell."[†] In accordance with this idea of art as saviour of the artist, Anthony Storr in his book *The Dynamics of Creation*[‡] cites an impressive number of cases where writers and artists maintain that their work is what kept them from falling over the edge. However, as with most correlational thinking and generalizations from subjective personal experiences, the popular conclusion does not necessarily follow from the data.

.

The first and most obvious criticism of this idea of art as therapy is that it is far from a universal description of the experience of creating art. Writers in particular are quick to describe what writing is like for them, and they certainly don't all agree with Artaud, although some do.

.

"Let's face it, writing is hell," William Styron has said.[§] Or as Jessamyn West put it: "Writing is so difficult that I feel that writers, having had their hell on earth, will escape all punishment hereafter."[**]

---

[*] Dryden, Absalom and Achitophel, I.

[†] Antonin Artaud "Van Gogh, the Man Suicided by Society (1947) in *Antonin Artaud Selected Writings* edited by Susan Sontag (Berkeley and Los Angeles: University of California Press, 1988) p. 497

[‡] Anthony Storr *The Dynamics Of Creation* (London: Secker & Warburg, 1972.)

[§] A comment made to George Plimpton in a *Paris Review* interview eventually published in the first volume of *Writers At Work*.

[**] In the first chapter of *To See The Dream*, Avon Books (1974).

And Red Smith, more ironically: "There is nothing to writing. All you do is sit down at a typewriter and open a vein."* On the other hand, many writers insist that writing is what keeps them sane, and find it hard to imagine how anyone who doesn't have writing as an outlet remains functional.

.

So, albeit without any real scientific tabulation to back this up (only the reading of hundreds of interviews with writers!), I'd say it's about a fifty/fifty split between the "it keeps me sane" group and the "it makes me crazy" contingent.†

.

The second, less obvious, criticism of the idea of art as therapy is that those who maintain that their art is what keeps them out of the loony bin may very well be self-deluded. Everyone knows people who maintain that daily doses of homeopathic medicines or drinking seven cups of herbal tea or never stepping on the cracks in the sidewalk is what keeps them healthy. Human beings have evolved to assume causal relationships (generally a good thing in terms of natural selection), and thus are innately predisposed to link together coincidental but fundamentally unrelated behaviours and outcomes.‡

.

A third criticism is that work—or even just being busy—is a well-established way of relieving not just anxiety and depression but even physical discomfort. Someone distracted by engagement in a task—and it need not be a creative one—is less likely to attend to the emotional demons screaming in her head.

.

A final criticism (although more a remark that might even be construed as supporting the idea of art as therapy) is that venting one's emotions is almost universally accepted as therapeutic—if it is done in a way that does not increase stress in one's life. Your spouse really pisses you off and you write a vicious, nasty poem about him or her. You feel better and you will *be* better—*as long as you then destroy the poem.*§

.

---

* Quoted in *In No More Rejections: 50 Secrets to Writing A Manuscript That Sells* by Alice Orr.
† My own experience, for what it's worth, is also almost fifty/fifty. On good days, writing is a joy. On bad days, it is only slightly better than having a root-canal done.
‡ Psychologists call this "superstitious behaviour". Should I happen to make a difficult pool shot after even facetiously calling on the ghost of my dead brother, I'll be more likely to call on bro's ghost again when confronted with a difficult shot.
§ Sylvia Plath, take note!

But doing art is not like that for several reasons. First, your spouse will get to read the poem (because art is a public activity) and that will only increase the stress in your life. Furthermore, in art the work is not merely an expression of the creator's feelings—as far too many would-be poets seem to think. If it is art, it is a creation of something that evokes some magical feelings in the reader. The idea of art *as no more than* self-expression is juvenile and naïve, as T. S. Elliot knew: "Poetry is not a turning loose of emotion, but an escape from emotion, it is not the expression of personality, but an escape from personality. But, of course, only those who have a distinct personality and deep, painful emotions know what it means to want to escape from these things."*

Of course, given the ingenuous willingness of psychologists and the public to accept just about anything under the umbrella of "alternative medicine", art therapy is now widely practised. Insofar as writing vicious verse or drawing one's demons is an effective way of venting one's emotions, it may indeed even be somewhat therapeutic. However, the efficacy of art therapy for the inartistic is beside the point. As far as one can tell, the therapeutic value of practising one's art *for the creative artist* is far from established. That some artists find solace and refuge from the turmoil in their lives and in their minds by creating is certainly true. But is this different from the solace and refuge everyone finds in work, in any busyness that distracts from one's grief and pain?

And one final possibility worth consideration is that creative people are 'addicted' to their creative work. The relief from withdrawal, the sudden high, of finally getting one's drug could easily be misconstrued as therapeutic. The heroin addict with shakes and chills is 'cured' when he makes his connection and injects that good juice in his veins. Monomaniacal obsession with one's work is one of the characteristics of creative individuals. It is thus hardly surprising if such individuals maintained that doing their work kept them sane. Alcoholics remain sane as long as they have access to booze. "I don't have a drinking problem; I only have a problem if I can't get a drink." The madness of DTs is prevented or 'cured') by getting that drink.

Before leaving this topic, something need be said about scientists—who have been neglected in this sub-section. They rarely describe doing their work as "hell", although they do sometimes describe

---

* From T.S. Elliot's *Selected Essays*.

being "tormented by a problem". The scientist's sleeplessness, irritability, and other neurotic behaviour patterns manifested while wrestling with a problem (and the subsequent relief and return to stability after solving it) imply that the creative act for the consistently creative individual in art or science is not so much therapeutic as a physical need—something I've previously offered as an explanation for the amorality of creative individuals at work. It is not surprising that confusion and ambiguity surround 'the fix'.

.

Leonard Cohen: "There is no cure for love." There probably isn't one for creativity either. But then who *really* wants one for either?

CASE STUDIES: LORD BYRON, AUGUSTA ADA BYRON

Lord Byron, the great romantic poet. Ada Byron, his daughter and the mother of modern computers. Artist and scientist.

By all accounts Byron could be a real bastard and by some accounts his daughter, Ada, a real bitch.* One of Byron's famous lovers, Lady Caroline Lamb, described him as "mad, bad, and dangerous to know." Ada was not nearly as mad and bad as her dad, so the above 'bitch' label is quite unfair and a cheap shot for rhetorical purposes. However, she was a drug and gambling addict, and while not as indiscriminately promiscuous as her father, did warm more than a few beds and leave behind a few (often eminent) embittered lovers, and allegedly could be hard as nails with those who questioned her intellect. However, as I've already suggested we tend to expect too much of creative people. Why on earth should they be any nicer than their less illustrious neighbours? But it is worth asking the question whether they sometimes seem in general to be *less* nice.

Lord Byron and his daughter, Ada, Countess of Lovelace, are an interesting contrast for several reasons. Byron was the 'mad' poet; his daughter was the 'mad' scientist. They both have been 'diagnosed', albeit of course *ex post facto,* as suffering from bipolar disorder. Since their only real connection was genetic (for Ada had virtually no contact with her father after the age of one month), the similarity in their lives, despite very different domains of endeavour, has been offered as evidence of the importance of genetic inheritance in creativity, with environment playing the secondary role of channelling that creativity.

George Gordon Noel, Lord Byron, was only figuratively speaking a bastard. He was born in wedlock to John Byron and his wife Catherine (nee Gordon) on January 22, 1788. His parents' lineage was not auspicious in terms of mental health. Insanity, murderous behaviour, and suicide were common among his ancestors. Several of his mother's relatives were convicted and hanged for murder. His uncle, on his father's side, was known as "The Wicked Lord" for his notorious and immoral actions.

---

* A rhetorical indulgence, I hope to be forgiven for after the reader reads on.

The poet spent the first years of his life in England and Scotland with his mother, while his father hid from his creditors in France—dying (probably by his own hand) when Lord Byron was only three. When the poet was ten, his uncle, the "Wicked Lord" died, and Byron and his mother moved to Nottingham, near what was left of the Byron estate.*

Byron was born with a malformation of his foot, which caused him to limp. Children then, as now (despite all the political correcting they are given) are naively cruel. Young Byron was to his peers at school just the gimpy kid who spent much of his time alone reading and dreaming about Roman military history.

But he also early showed signs of the raging libido for which he remains notorious. He fell in love (or lust) with a neighbour and cousin named Mary Ann Chaworth. But she married another and in 1805 he started studying at Trinity College, Cambridge. It was there he again fell in love (or lust), this time with a fifteen year old choirboy. When his young love died suddenly, Byron wrote a series of elegies, *Thyrza*, in his memory—changing the pronouns to disguise the homosexual nature of their relationship.

At twenty he was already publishing his poems, poems inspired by his already numerous passions, and including poems considered shocking to the genteel audience for poetry of that time. His academic 'career' was a joke. He had no use for Cambridge, except as a home base for his literary and sexual exploits. In 1809 he set off with several friends on a European tour that was to last three years and end in the Middle East. His extreme sexual exploits and literary productivity increased exponentially during this trip and throughout the rest of his life.

The sex life of Lord Byron would make an epic pornographic movie, and there are many scholarly (wink, wink, nod) biographical tomes devoted to his life. What is relevant here is the relationship of his personal and literary life, the paradox of it, and his commitment to high moral principles that so contrasted with his personal immorality. Byron is the archetypal romantic rebel. His personal sexual life was so far beyond the pale that eventually he was effectively exiled from his homeland. (Yet his books were best-sellers back home.) Before his exile he served in the House of Lords and defended Luddites (those

---

* To pay huge gambling debts The "Wicked Lord" had felled the estate's magnificent oak forest, sold 2,700 head of deer and leased for a ridiculously low sum rights to land that was rich with coal.

opposed to 'modern' technology as degrading), as well as Roman Catholics and social reformers. However, he was almost certainly guilty of an incestuous affair with his half-sister, and definitely "mad and bad" when it came to most interpersonal relationships. But, on the other hand, he spent his own money to support the insurgent movement for Greek independence from the Ottoman Empire and actually, despite any military experience or competence, joined their army to fight for what he felt a noble cause! It was while waiting to sail to war that he fell ill, and through the incompetence of the attending doctors died from a 'bleeding cure' in 1824.

.

What to make of such a man? He loved animals, at various times having as pets a fox, monkeys, a parrot, cats, an eagle, a crow, a falcon, peacocks, guinea hens, an Egyptian crane, a badger, geese, and a heron—and a bear he kept for a time at Trinity College in protest against the rules against students having dogs! And, most famously, a Newfoundlander dog named Boatswain, for whom he wrote a famous epitaph.

.

He also loved, usually carnally, with very little more discrimination, both men and women. That most of these love affairs ended tragically and bitterly is not surprising. He was as prone to fits of violence as he was to fits of unbridled loving passion—and as to fits of creativity. But by all accounts he was when loving, *very* loving. So may have been "mad and bad" but he was also a great poet and sometimes a great person.

.

And what of his daughter, Ada, who he never really knew? Was she also "mad and bad", an apple not falling far from the tree? By some accounts she was utterly charming, but by others she was his female equivalent in terms of the good, the bad, and the ugly.

.

Well, to deal with these attributions in reverse order...

.

She definitely was not ugly! Like Lord Byron, more so actually, she was very attractive. Extant portraits confirm her beauty. And she took hedonistic advantage of her charms to seduce some quite exceptional men. Among those smart enough to have allegedly been allowed to share her bed were such eminences as Charles Babbage (mathematician and one of the first computer scientists), Sir David Brewster (the originator of the kaleidoscope), Charles Wheatstone (inventor of the stereoscope, concertina, and an early form of the microphone), Charles Dickens (the great novelist who needs no introduction), and Michael Faraday (the famous physicist and chemist who contributed so much to our understanding of electromagnetism). I really don't know if there were any stable boys from the racetracks she frequented that gained access to her bedchambers, but it seems that generally she was quite selective for intellectual eminence in her lovers.

.

Was she bad? Well, while she didn't compete in the same league as her father, she was a long throw from a bourgeois paragon of stability and conventional virtue. She lived fast and hard, with a complex and—by the standards of the time—promiscuous and scandalous sex life. She had married at the age of nineteen William King, a few years later to become The Earl of Lovelace, but fidelity wasn't her strong suit as the list of her alleged lovers indicates. She also loved to gamble, and unfortunately wasn't good at it despite her mathematical sophistication, having to pawn her jewellery to pay for her gambling debts and still accumulating over two thousand British pounds in debts before her death in 1852. She was also a serious drinker and sometime user of opium—once expressing the intention to write a 'scientific' treatise on the effects of opium and wine based on her own personal experiences. She was reported to have extreme mood swings, as one never knew whether she would be charming and witty or acerbic and vitriolic. Ada was no genteel noblewoman, content to sit in her garden and crochet doilies, demurely nodding acquiescence to any inane pronouncements of visiting male nobility. So she was a very bad girl by the standards of the time, and would probably still be considered so even now.

.

And was she good? No. She was far more than good—intellectually. She was indisputably brilliant—and brilliant within a culture that very much distrusted intelligence and independence, never mind brilliance, in its women. Furthermore, her brilliance was in the domain least acceptable for women of the time: science and mathematics and logic. Her contribution to the theoretical basis of computer science is universally acknowledged. The whole fundamental concept of a computer ("The Analytic Engine") was the result of her collaboration with colleague (and alleged lover) Charles Babbage. Ada's extensive comments in her translation of Babbage's seminal paper include an algorithm to compute Bernoulli numbers. This is said to be the very first computer program and so attributes to Ada Lovelace the honour of being the first computer programmer. (So much for the stereotype of computer programmers as being pimply guys!) The AI computer programming language ADA is named in honour of her.

.

Annabella (nee Milbanke) Byron, Ada's mother, had already had more than enough of mad and bad Lord Byron when their daughter was born. Five weeks after Ada's birth, Lady Byron separated from her husband and was granted sole custody of their daughter. Byron left for Italy and never saw his daughter again. Annabella tried to raise Ada to be anything but like her father. She gave her daughter all the advantages that come with nobility: lessons in music and gymnastics and equestrianism. Ada was sent to attend the best theatrical performances, finest concerts, and most elegant parties. And she apparently took to it all with great enthusiasm, becoming an accomplished amateur musician and athlete. Her love of the arts also grew, but this was less encouraged by her mother, who feared above all else that Ada would follow in her father's footsteps and become a poet. Poet had come to be equated with mental instability in her mother's mind. In an attempt to divert Ada's voracious appetite for intellectual and creative endeavours, Annabella encouraged her daughter to study science and mathematics. At the age of seventeen, Ada met a woman, Mary Somerville, who had just written a book on mathematical astronomy and who was to become Ada's mentor in mathematical studies.

.

Ada, like many mathematicians, seemed to find common ground between math and music, and, despite her mother's acquired aversion to poetry, which Anna had tried to inculcate in her daughter, Ada also retained a deep love for the literary arts. She was a Renaissance woman in many ways, accomplished in many domains and a brilliant intellect. It is no wonder that many feminists want to set her up as role model, albeit usually playing down her foibles and fondness for men.

.

So does the apple ever fall far from the tree? Well, the evidence in this case clearly points to nature, not nurture, being convicted—or credited. Ada never knew her father and so his 'nurturing' influence had to be virtually non-existent, yet the similarities between dad and daughter are hard to ignore. The approach her mother took in raising her, and its results, offer support for the idea of Nature dealing the cards and Nurture deciding how they are played.

.

As previously mentioned Byron's family history is often cited as evidence of the heritability of both genius and madness[*], an

_____

[*] Kay Jamison uses Byron and his family as a classic case of bipolar disorder heritability.

inextricably entangled inheritance. There is no question that Annabella believed that her daughter was born gifted and cursed with both, because of her genetic legacy from her father, and so Mother tried to channel her daughter's gifts away from the literary life which she feared would aggravate Ada's mental instability by its emphasis on emotion over reason. Instead mother encouraged an interest in science and mathematics, 'rational' outlets for her daughter's native intelligence and imagination. Unquestionably she was to a certain extent successful. While Ada never lost her interest in the arts, she did direct her energies toward the sciences and mathematics. And while the tinge of 'madness', the emotional lability and intensity, still manifested itself in her life, they were probably attenuated by her immersion in a creative environment that tended to value head above heart.

.

Lady Ada Lovelace, like her father, died young. Ironically, she, like her father, may have died of what was then considered a cure: the "bleeding cure", which was applied as a treatment for her uterine cancer. She died at the age of 36. Her father also died at the age of 36. She was buried beside him in the Byron family vault in Hucknall, Nottingham England: Artist and scientist, father and daughter, side by side for eternity.

OUR SENSORY CENSORS: OUR ALL TOO HUMAN LIMITS

"The limitations of our senses are the walls of our creative prison."
—Hippokrites

"...life has not been devised by morality: it wants *deception, it* lives *on deception..."*
—Friedrich Nietzsche *(Human, All Too Human)*

Bees' eyes and bats' ears. These are reminders of the biological limitations of our senses. Our receptors do indeed censor our experience of the world, but the limitations of our sensory organs are only the first line of defence against full experience of the world. Our brains take over the filtering of the information that gets past the receptors in our eyes, ears, skin and other organs. We do not *attend* to every neuronal input. To some extent, we voluntarily attend and ignore, but to a great extent the focus of our attention is not volitional.

No serious artist or scientist I've ever known would defend external censorship, but the sensory censors we all have to live with, be they the limitations of our receptors or of our brains, are perhaps even more maddening than some Boston puritanical bureaucrat banning a book. What painter wouldn't like to see her palette expanded to include the undreamed of colours a bee must see in her garden? Sorry, but human cones in the retina are only photosensitive to electromagnetic radiation in the range from about 400 to 700 millimicrons in wavelength. What composer wouldn't like to be able to hold the attention of his audience for eight hours of intricate musical invention? Sorry, the human attention span to complex auditory input is far shorter than that.* We can fight the human censors, maybe get the prudish little old lady on the censor board fired for moral depravity, but we can't fight biology.

---

* Beethoven's innovative Eroica Symphony, which is about 50 minutes in length, shocked his audience and pushed the limits of their attention span, for those used in symphonies by Mozart and Haydn were about half that length.

Or can we? Well if not fight it, then at least we can work with it and around it. Or use technology to overcome it.

.

I cannot walk out on my porch on a summer night and gaze up at the sky to savour the colourful beauty of distant nebulae. But powerful telescopes can look up yonder and 'see' what my retina can't, and computers can beautifully colourize what these telescopes see, and the images can be digitized and posted on the NASA website for my aesthetic delectation—not on my front porch, but on the computer monitor in my study.

.

The painter of a landscape cannot make the two dimensional surface of her canvas three-dimensional, but by understanding how we perceive three dimensions she can at least simulate the effect of true binocular, stereoscopic, perception.

.

The reason science has traditionally been the handmaiden of art, not its arch enemy, is that understanding how we perceive and attend allows the artist to work around the sensory censors. The example of perspective painting during the Renaissance is the classic example of this, where scientific and mathematical knowledge, as well a deep understanding of human perception, was considered prerequisite. Like the jujitsu master, artists turn the strength of their opponent, biological limitation, to their own advantage.

ARTISTIC VISIONS AND ILLUSIONS

Back in the year 2000 I visited a major retrospective exhibition at the Smithsonian's Hirshhorn Gallery of Contemporary Art in Washington D.C. entitled "Dali's Optical Illusions". Salvadore Dali, of melting clocks fame, was a surrealist who was especially fascinated with optical effects and illusions. He painted several famous anamorphic* religious works, and almost all of his work shows a profound understanding of the oddities of human perception, which he exploits to great effect. Some critics have gone so far as to suggest his art borders on clever kitsch because of the many optical illusions embedded in so much of it. This is unfair, for although I personally don't think he is the greatest of the surrealists and am annoyed that his antics and self-promotion have overshadowed more significant surrealists such as Magritte, Enrst, Tanguy and DeChirico, Dali's really serious study of perception has resulted in some great works of art.

.

But what struck me as I left the show was the perhaps obvious—but important and often unappreciated—insight that virtually *all* visual art is based on optical illusion. One of the jobs of the painter is to fool the eye of the perceiver, the appreciator. The most commonplace example is the technique of realists, who understand perspective drawing and know that the illusion of three-dimensionality can be created by the use of relative brightness, amount of detail, and the Gestalt principle of size constancy. This is a trivial observation, but when one considers the difference in our reaction to a simple optical illusion as opposed to a genuine work of art, it holds deeper implications. We are momentarily fascinated by an optical illusion, but with art, the illusion is not an end in itself, but rather a means to a deeper end.

.

It is in fiction that we often most profoundly apprehend real life. It is often in painting that we truly apprehend and appreciate the visual world. Strangely enough, often the illusion that the artist creates is experienced by the perceiver as more intensely real than the real world itself as usually experienced. Furthermore, the aesthetic

---

* An anamorphic painting is a distorted image stretched vertically or horizontally. It has to be viewed from an extreme angle in order for the image to be undistorted. Dali's famous Crucifixion has to be hung very high up for proper viewing.

experience is like a magic mirror: it reflects back on the real world, brilliantly illuminating it and making reality *more* real to us.

Consider representational painting as one example. When I first arrived in Canada, a man who was profoundly passionate about the arts befriended me. He was also very knowledgeable about Canadian artistic accomplishment and anxious to share his enthusiasms with this new immigrant. He introduced me to the work of the early twentieth-century Canadian painters known as The Group of Seven. It didn't surprise him that I was fascinated by their work—for he knew I loved the Northern Ontario bush, and they were painters of this landscape, the same landscape I would, whenever possible, escape to in my canoe.

This group of painters is extremely important in the history of Canadian art, for they were the first to break from the imported and imitative style of art that dominated what little art scene there was in Canada during the first decades of the century. In Europe Modern Art was being born with all its diverse radical schools and approaches to painting. Never mind the surrealists and futurists, Kandinsky was already producing the first abstracts; but Canadian artists weren't being influenced by, and importing, theses styles and approaches. They were truly provincial, imitating of all things the old school of pastoral landscape painting. Trying to capture the soul of the wild Canadian landscape in the tame pastoral style of a Constable is a really bad idea, and the Group of Seven realized this.

They weren't entirely up to speed with the contemporary, frenzied European art scene, either, but they knew the work of post-Impressionists such as Van Gogh, Gauguin and Cezanne, as well as, interestingly, the work of Japanese artists and the Art Nouveau graphic artists. From these rather unlikely influences, they created a new style appropriate to the rugged Canadian landscape. It would be silly to try to describe this style in words; images of their works can easily be found on The Internet. Suffice it to say, their landscapes of the boreal wilderness 'get it'—and they got me.

These artists got my attention. They '*caught* my attention'. This common phrase says much. As I've already said, our senses detect much more than we fully experience, for we have very selective attention—and it is a powerful censor of our experience. What is it that focuses the lens of our attention? Knowledge. Knowledge gained through art and through science. (The latter I will write of later.)

The works of The Group of Seven were for me a focusing lens that allowed me to see more clearly a part of the world I already loved feasting my eye upon. One of the artists associated with The Group of Seven is Tom Thomson. In some ways he was the founding father of the movement, but he died at age forty*, just before the group was actually formally formed. He worked at a commercial graphic design company with many of the artists who eventually coalesced into The Group of Seven. Like his colleagues, he was attracted to the wilderness several hundred kilometres miles north of Toronto, and spent more and more time in Algonquin Park, where he worked as a guide and fire-fighter when not painting. Algonquin Park is a 7,725 square kilometre uninhabited nature preserve in the mixed boreal forest region of mid-North Ontario containing 2,400 lakes connected by 1,200 kilometres of streams and rivers. Even now, only one paved road passes through a small part in its southern portion; in Tom Thomson's time is was truly pure wilderness. Like Thomson, although more than half a century later, I used to escape Toronto to canoe trips in this gorgeous landscape where encountering another human being after the first portage was unlikely.

When, one wintry urban evening, I first viewed quality reproductions of Thomson's paintings in one of my friend's art books, I immediately felt I was seeing Algonquin Park even more clearly, more *noumenally*, than when I was actually there. And then when spring arrived and I again ventured into the park, the landscape had been transformed. Of course, it was not really the landscape that was transformed: it was my eye—or rather more accurately—the part of my brain that filters and revises and *appreciates* the data coming in along my optic nerves.

This is the two-fold magic of art: its ability to focus our attention on itself *and* to increase our ability to focus on the real world. Art helps us to see. To speak of artists as visionaries is apt. They transform their visions into illusions, which then in turn enhance our vision. This double transmutation is a wondrous mystery.

---

* He died under mysterious circumstances on a solo canoe trip.

SCIENTIFIC VISIONS AND DELUSIONS

A delusion is conventionally distinguished from an illusion in that an illusion is an erroneous perception and a delusion is an erroneous cognition or belief. But 'erroneous' is probably a misleading, and unnecessarily pejorative, word in the current context; for just as I've tried to show how illusion is central to arriving at deep truth in art, I would now argue that delusion is essential to arriving at deep truth in science.

.

Because science has been so very successful as a methodology for apprehending the world, and because science is based on observation, we rarely reconsider the fundamental principles on which it is based. Before science had become so well established, many philosophers did question the cornerstone assumptions of scientific interpretation. For example, David Hume, a philosopher who is often considered an empiricist, pointed out that even causality could very well be a cognitive delusion hard-wired into our biology. Temporal and spatial contiguity, as in B happens near A in space and shortly after A, is hardly sufficient justification for endorsing the incredibly abstract concept of causality.* Correlational thinking is obviously flawed by the unjustifiable assumption of relationship equalling causation, but even the best designed and controlled experiments have a variation of this assumption behind them.

.

Most scientists hold sacrosanct a set of beliefs that are at bottom mere intuitions (albeit now widely shared intuitions), and may very well be delusions. I am no epistemologist armed with all the sophisticated tools required to dissect these beliefs, but it has been done. My point here is that at least some of these beliefs *may be* delusional, and in some cases there is good reason to believe they are; and if not demonstrably delusions, at least not fully warranted beliefs. Yet these 'delusions' allow the scientist to help us apprehend the world in a deeply satisfying way—just as artists use illusions to do the same thing.

.

---

* "We have no other notion of cause and effect, but that of certain objects, which have been always conjoin'd [sic] together, and which in all past instances have been found inseparable. We cannot penetrate into the reason of the conjunction. We only observe the thing itself, and always find that from the constant conjunction the objects acquire an union in the imagination." (David Hume in *A Treatise of Human Nature.)*

Causality is only one of this set of beliefs, but is a crucial and paradoxical one. The creative scientist must swear allegiance to this belief if he is to practice his craft. But without wandering off into an age-old philosophical quagmire, the principle of causality at the heart of science seems to lead inexorably to determinism, but the universal, *empirical* experience of freedom of the will is a clear refutation of determinism. Surely if science is based on empirical evidence, then such universal empirical experience can't just be ignored! But it has to be, or the enterprise of science would have to shut down all its labs and close up shop. So, for practical purposes, it is.

.

Another critical belief is faith that our senses are to be trusted, that at least in some ways they are delivering meaningful data from the world outside our skin. Again philosophers such as Berkeley and Hume have pointed out the insupportable nature of this assumption. When I teach a unit on human perception, I like to describe in fast, flickering detail what is, according to science itself, happening when one sees something.

.

I'm hungry, having skipped breakfast to make it to my 8:30 a.m. class on time. I'm delighted to see a student has placed a delicious looking bright red apple on my lectern. I eye it. The electromagnetic radiation (which would be in the range of 600 to 700 nanometres, millimicrons, in wavelength) not absorbed by the skin of the apple is reflected back at my eye. This light passes through my cornea, which bends it enough to invert the image 'information' it is 'carrying'. Then it makes its way through the fluid (the aqueous humour) behind the cornea. Those photons not blocked off by the pigmented circular muscle called the iris continue on to pass through the hole in my iris called the pupil and enter my crystalline lens. My lens changes shape because of my ciliary muscle tugging at its edges in response to my conscious intention to see this apple thing clearly. These changes in lens shape bend the incoming light just enough as to project a clear inverted image on my retina. The light then continues on through the clear jelly (my vitreous humour) that is maintaining the spherical shape of my eyeball. When it arrives at the back lining of the interior of my eye (the retina) it still has to wend its way through a tangle of neuronal 'wires' before entering the receptor cells called cones that cluster in the foveal, central, area of my retina. These cones contain photosensitive chemicals. Some are particularly responsive to electromagnetism of the same frequency as that of the incoming light. (I don't lack these particular cones so am not colour blind.) They 'bleach' out and stimulate adjoining cells called bipolar cells which in turn trigger ganglion cells, the axons of which bundle together and leave the eye. This bundle is called the optic nerve. These nerve bundles meet at the optic chiasma where half the axons from each nerve cross over to the opposite side of the brain. The optic tracts eventually project onto an area in the brain's occipital lobes. Here in this visual cortex and in secondary areas, many more neurons are set into action, with sodium and potassium flowing back and forth through the cell membranes and neurochemicals being dumped into synapses. And lo, I have the experience of seeing the apple!

.

Lord in Heaven! C'mon! What does all this tell you about my pleasant subjective experience of seeing this lovely red apple just when I most desired a bit of nutrition before setting out to describe visual perception? And what does my experience have to do with the reality of the apple? It has no colour. Colour is merely a subjective experience resulting from the above described series of events which result in chemical changes in the neurotransmitters in my brain. Remove me from the room, and the apple has no colour. There is no colour in the universe. What colour are wavelengths that don't affect the photosensitive chemicals in my cones? What colour is microwave radiation? What colour are gamma waves?

.

And this apple. What is its function? To ease my hunger? Actually it has no function independent of me either. All 'function' is 'in' the human brain, like colour, a result of consciousness which is no more nor less than an emergent property of the electrochemical doings of the organic soup slopping around in our craniums. Without my hungry presence at this physical event, the apple has no 'function'.

.

Of course all of the above is based on accepting the findings of science as valid, and here we find the serpent eating its own tail—or tale! I'm using science here to discredit trust in observation, but observation is the basis of science.

.

Philosophers, bless their anachronistic souls, are working on these problems, and in the meantime the average scientist and the average person does not question causation or the validity of perception. Implicit acceptance of both allows us to understand the world, and artists can and do use this understanding to create the illusions that help us understand the world in a different way.

.

But to return to Algonquin Park—this time with an example of how it was science that deepened and enriched my experience of its landscape. One of Tom Thomson's paintings is entitled "The Jack Pine". Looking at it I realized that I didn't know jack about pines—or for that matter any of the trees that defined the landscape I so loved. So one winter I bought a book on the indigenous trees of Canada and studied it. Returning to Algonquin in the spring, the landscape had again metamorphosed. What once was an amorphous wall of trees was now thrown up in high relief. Over there was a stand of birch, and a bit further along the shore white spruce Near the creek mouth were some tamaracks and black spruce. And on that small island a jack pine very reminiscent of Thomson's painting.

.

It is amazing that something as simple as scientific taxonomy, the mere naming of things, can so quickly throw into high relief what was before so flat and undifferentiated. And anyone who has travelled with a talkative naturalist knows that more sophisticated scientific knowledge can sharpen even more one's perceptions. You first recognize the jack pine, and then your vision of it becomes even more detailed and rich when you realize why it is growing where it is growing, why it has the shape it has, and how its roots are spread out beneath the shallow soil.

.

I know that Tom Thomson had a naturalist's knowledge of the landscape he captured so well in his paintings, and I don't think it unreasonable to assume that this knowledge enriched his work, even if it was not explicit in it. As another great painter, Andrew Wyeth, once remarked: "It's not what you put in but what you leave out that counts." But of course for this to be true, what you are leaving out first has to be inside—you.

MIXING OUR ALREADY DISORDERED SENSES: SYNESTHESIA

Synesthesia refers to a condition where the input to one sense triggers a consistent perception in another sensory modality. Numbers or letters or musical keys may have intense colour experiences associated with them, which is one of the commonest manifestations. But the input can be via any sensory modality and the associated perception in any other. For example it can be tactile: some synesthetes taste shapes. If this seems difficult to imagine, try to extend those common metaphorical examples we are all familiar with into a more direct perceptual experience. Why is one musical genre called "the blues"? Why do we refer to some musical compositions as "black"? Why do we call the taste of certain cheeses "sharp"? Why do we call some bright and highly saturated colours "loud"? Or try the following little experiment.

Imagine in your mind's eye an inkblot and a jagged piece of glass. Your task is to guess what they are called in an obscure language— which would be an impossible task except that you're told that one of them is a "bouba" and the other a "kiki". Well, which is which?

One study indicated that 98 percent of people asked this question answered in the same way, while people with known damage to a particular part of the brain (the angular gyrus) seemed to pick randomly.* But for most of us the answer seems obvious, although we may have some difficulty in articulating why. With this example, one immediately thinks of onomatopoeia and how poets use the implied 'shape' of sounds and phrases to express and describe. This can only be effective because we are almost all, at least to some extent, synesthetes. Composers too use what they call "colour" in their music. Beethoven always referred to the key of B Minor as the "black key" and Sibelius insisted G Major was green. Virtuoso rock guitarist Jimi Hendrix considered the chord E7#9 to be purple, and it is central to his famous song "Purple Haze". Even the timbre of some musical instruments like the clarinet are said to have "bright colouration", or sometimes are alternatively described in terms of tactile sensations, such as "warm". (Jazz as a genre is often described as either "cool" or "hot".)

---

* This study is cited in an April 2003 *Scientific American* article entitled "Hearing Colors, Tasting Shapes". Of course, it should go without saying that any readers who think the inkblot is the "kiki" are not necessarily brain-damaged, but they are unusual in their response.

There is evidence that many artists (as well, as mathematicians and scientists) were profoundly synesthesiac. Several notable poets, including Baudelaire and Rimbaud, were synesthetes, and Feynman saw algebraic symbols of Bessel functions in colour. This may be no coincidence, for there is now some evidence that full blown synesthesia is more common in creative individuals. Furthermore, it is not preposterous to see metaphor as an extension of this mixing of perceptual experience, for what is metaphor if not finding associations between different domains, even if not always distinctly sensory domains?

The synesthesia phenomenon was first remarked on in a scientific paper in *Nature* back in 1880 by a man whose name seems to surface again and again in any examination of creativity: Francis Galton. Still, the phenomenon was for a long time considered a mere curiosity; it is only recently that some researchers are suggesting it may offer deep insights into how human perception works.

Synesthesia is more common in the young. One theory, proposed by D. Maurer back in 1993 and elaborated by Simon Baron-Cohen, is that this mixing of sensory modalities is universal in the very young, but as our brains become more specialized it fades away in most people, usually by puberty. The so-called Neonatal Synesthesia Hypothesis is that our sense modalities are undifferentiated at birth and only become independent by around four months of age. A less radical version of this hypothesis simply says that objects are simultaneously recognized in more than one sense modality in the very young. This latter hypothesis certainly has strong empirical support from ingenious experiments with infants.

What are the implications of this? *Homo sapiens* is an exceptional species in that humans are still 'prenatal' when they are neonates; and remain childlike and playful all their lives*—something which seems especially true of creative individuals. A strong argument could be made, and perhaps I have already inadvertently been making it, for the creative individual being 'a case of arrested development'. In that case, synesthesia being more common among the creative should come as no surprise.

---

* It has been suggested that our love of dogs is partially explained by the fact they, unlike their wolf ancestors, don't 'grow up', but instead remain playful all their lives.

There is another sense (pun intended) in which synesthesia is important in understanding creativity. If creativity requires 'seeing' connections between things that are not obvious, which is what metaphor is about, then synesthesia is itself a metaphor for the creative process. There is something different in the neural wiring of the synesthete, some lack of the specialization and compartmentalization typical of adult brains. Even for those who are not literally synesthetes, such sloppy wiring, such short-circuiting, and such flexibility, is essential to creative insight.

MIXING OUR MANY MANGLED METAPHORS: MULTI-MEDIA ART

A more direct method of 'mixing' sense modalities is to combine them in the creation of an art work. Decades ago, roughly around the time of the advent of personal computers with graphical interfaces, the term 'multi-media' became a common buzz word for art and entertainment that did not limit itself to one medium or one sense modality.

.

Of course multi-media art is as old as art itself, and until around the time of The Renaissance it was the norm. Festival and theatre have always been multi-media events, feasts for both eye and ear and often the other senses as well.

.

However, after book publishing became financially feasible, after visual art became separated from a larger context, and after absolute music was developed and its dominant performance venues shifted from church and theatre to concert halls, and artworks of a purely literary, visual or aural nature became common.

.

People read books, silently and to themselves, rather than listened to stories in verse (or prose). Poetry was not declaimed. And those without good internal mental ears to hear the sound of it as their eyes scanned the pages found it perplexing rather than musical.

.

People went to art museums to silently examine artworks isolated from any other explanatory context. Beautiful altar pieces were not part of the whole complex, 'multi-media' experience of attending a church service in a grand cathedral. They were just *objets d'art* to be stared at against a neutral white background and commented on in whispers.

.

People attended concerts in dark halls, lest any visual activity distract them from concentrating on the purely aural input they were to focus on. And more often than not, even words to accompany the music were *verboten* and considered a distraction.*

.

Such compartmentalization is no longer as wide-spread as it was in the recent past. Oral poetry is making a come-back with poetry slams and the acceptance of 'dub' or improvisational poetry as legitimate

---

* Opera and ballet obviously are exceptions, for they remained multi-media.

genres. Contemporary visual artists are more likely to do installations and performance pieces than create a painting for a museum wall. Pop music was transformed by the advent of music videos, and pop concert performances are unquestionably more spectacle than a purely aural experience. (Meanwhile, the audience for traditional concert music performance is becoming disturbingly small.)

Some of these changes are to be applauded, and some may be disturbing, but what is relevant here is the effects this has—and what demands this makes—on the creative individual caught in this resurgence of multi-media art. Not every writer or painter or composer wants—or feels competent—to step outside study or studio and wrestle with new media.

There are many important aesthetic issues associated with multi-media art. One is the question of whether or not more really is more, or rather may sometimes be less. The more our senses are engaged, the less our imagination is. How often do we hear people say they don't want to see a movie based on a book they loved, because it will almost inevitably be a disappointment, and there is also a risk that the visual images on the screen will replace those in their mind's eye? I may not be typical, but I'm certainly not alone, in often finding radio dramas more engaging than film or live theatre. And much as I admire the film *Fantasia*, I feel I've lost something in having the visions in my head once associated with some of the great concert music used in the film replaced by Disney's animations set to this music.[*]

Another question associated with multi-media art is the relative importance of each component part. The soundtrack of a film can make or break a film, but usually it is playing second fiddle (ahem) to the visual component.[†] And we speak of Shakespeare as a great *writer*, when of course he was actually the *script* writer for what was in his time what movies are for ours. Having humanities students only *read* his plays is in some ways equivalent to having students in a film studies course only read screenplay scripts.

---

[*] Stravinsky's publisher sued the Disney Company for unauthorized use of excerpts from his *Rites of Spring*. Stravinsky, after watching a screening of *Fantasia*, said he found the musical performance "execrable".
[†] 'Abstract' films for which Philip Glass wrote the soundtrack, such as the Qatsi trilogy, are an exception.

But the question that is most relevant to creativity per se is whether having to create using more than one medium is asking way too much of the artist. Exceptional talent in one domain rarely is associated with exceptional talent in others. The implication of this is that highly creative multi-media artists are going to be very rare, so the creation of most multi-media art will involve collaboration. And collaboration means that the individual vision or voice can not shape the whole. If creativity is uncommon even in individuals, it is far, far rarer among groups.

It might help to better understand this problem by considering three examples: so-called 'pop music', film (or movies), and opera. The first two are probably the most dominant art forms in our culture. The last is of historical interest and helps put the other two in perspective.

Pop Music

If you ask the average person what kind of music, or which musical artists, they like, you're unlikely to receive as your answer Mister Mozart or Johnny Bach. These days the word 'music' for most people means so-called 'pop music'. If their musical taste is at all developed, they won't respond by saying "easy listening", "soft rock", or name some group getting the most AM air play, and their preferences will probably be very far from what is really popular. (It hardly need be said that music lovers now are every bit as knowledgeable and sophisticated in their taste as those of Mozart's time, even if the majority of them have no interest in traditional concert music from the last three centuries.) Nevertheless, whatever genres or groups or individual artists they do say they like, the music will almost always be music accompanied by lyrics.* This is not something unique to our time. In fact, historically music without lyrics is the exception, not the rule. More about this shortly.

The creation of music with lyrics that does justice to both sound and word requires both musical and literary talent. This is asking a lot of the artist. I always get a gagging feeling in my throat when I hear someone introduced as a 'singer-songwriter', for only too often these individuals may have some talent in one of the two domains, but are totally incompetent in the other and their creations embarrassing. Fortunately, in some cases one talent carries the other—and occasionally an individual has both.

---

* Jazz aficionados excepted.

Incidentally, it seems to me that a more accurate phrase than 'singer-songwriter' would be something like 'composer-lyric writer'. What distinguishes these artists is the ability to write both music and the accompanying lyrics. Certainly, in some cases they are also capable of singing, but often the 'covers' of their creations by other performers are far superior to their own presentations of the work. Just as great writers may be horrible, stammering public readers of their own work, just as Shakespeare may have been a lousy actor, such artists as those about to be considered may not be the best performers of their own work. (In fact one has a voice like aural gravel, another one a distressingly nasal twang, and the last has a lisp!)[*]

So consider these three interesting and representative cases: Leonard Cohen, Bob Dylan and Tom Waits. These are real first-rate 'singer-songwriters' with which most people will have some acquaintance. Disentangling the lyric text from the music in each case is revealing. The simplest way to do this is to read the lyrics as poetry, just as words on a page, while trying to keep the associated music from playing along in one's head.

Leonard Cohen is first and foremost a poet, although he has also published two fine novels. However, his fame outside of literary circles is the result of his transformation from poet to singer-songwriter. As a writer I would rate him as one of the great romantic—or even metaphysical—poets of the Twentieth Century. (Even those who disagree with this hyperbolic evaluation, have to admit that the boy *can* write.) As a singer, he is—well let's just say he ain't Luciano Pavarotti. As a young poet in Montreal, he decided to spice up his poetry readings by accompanying his reading with guitar strumming. His early poetry is well suited to musical accompaniment, for he uses traditional troubadour and Meistersinger devices, and he *can* write a tune. In North America, poets usually have no hope of making a living at their art or attracting a large audience, but when Cohen started presenting himself as a singer/songwriter, he took on a role where there was at least some faint hope of modest fame and income. It turned out to be more than modest.

Many of Cohen's early songs were nothing more than his poems put to music. "Suzanne" is probably the best known of these.[†] It appears

---

[*] Mind you, I still love to listen to them.
[†] Its fame is more because of Judy Collin's cover of it than Cohen's own performance of it.

in his 1966 poetry collection *Parasites of Heaven*. If one looks just at the lyrics of these early songs there is no doubt as to his genius as a writer, and if one listens to the music without attending to the lyrics, there is evidence of a more modest, but still substantial talent, for creating memorable tunes. Later in his career, however, he did what most singer-songwriters do: writing the music and the lyrics as a single activity. The lyrics are no longer quite as capable of standing on their own, but the music has become more complex and interesting. And the melding of the musical ideas with the lyrics is more seamless.

.

Bob Dylan is a very different case. I am constantly told by people who have never in their life cracked open a poetry book that he is a great poet. I even recently heard an interview with some academic who is making a career out of comparing him to Shakespeare and John Donne.* I am a fan of Bob Dylan, and consider his diverse musical creations of enduring value, but he is a lyricist—not a poet. In my youth, I got my hands on a book of his lyrics. I still remember my profound disappointment in discovering that his lyrics stripped of the music were appallingly bad poetry. How could the same words that when integrated into his songs had such an aesthetic effect seem so downright silly and ill-composed when made to lie naked on the page? The explanation is that Dylan is a man gifted with musical lyricism and pure musicality—an extremely rare combination. Unlike Cohen, he didn't start his artistic journey as a poet and gradually meld it with musical composition. He is far more musically savvy, and his wildly careening travels through musical genres evidence a truly unique gift. That this gift includes the ability to meld words with musical ideas and imbue them with significance they could not have managed alone on the page is remarkable. To call him a poet is not only wrong: it is to do him a disservice. His gift is even more unusual.

.

And then there is Tom Waits. Waits *is* a poet. This is not to say he had a literary career, like Cohen had, before turning to music. It is to say that his lyrics, unlike Dylan's, do stand on their own two (metrical?) feet; it is to say that as nothing more than dark marks on white paper in a sound-proofed chamber they still *sing*. If Waits had never recorded a single song, but had someone had the acumen to publish his lyrics, he would be a literary star. So in some ways, Waits creative range is even greater than that of Bob Dylan.

.

---

* I assume this fellow is either joking or is an ageing hippie with tenure—and residual effects from drug usage in the sixties.

These three exceptionally creative artists illustrate different successful solutions to the problem of creating art that requires distinct abilities—the verbal and the musical. What about those working with media that engages even more of our senses?

Films (And Movies)

There is a substantial connotative distinction between the words 'film' and 'movie'. We tend to call a cinematic creation a 'film' to imply it is an art work; to call it a 'movie' suggests it is a mere entertainment. Without discussing the problematic validity of distinguishing between art and entertainment, I think it is fair to say that some cinematic works reflect a single artistic vision and many are a dog's breakfast.

There is no question that Hollywood has produced some major works of art, but it is associated with an approach to making movies (or films) that is contrary to every established principle of creativity. Great art is not created by a committee. Hardly anything great is created by a committee. What's the old joke? "What is a camel?" "It's a race horse designed by a committee."* Hollywood is notorious for going through script writers like the fellow hiring migrant day workers from the back of his pickup truck every morning. It is no wonder there is so rarely any real coherence to the scripts.

One major reason the ratio of cinematic chaff to wheat is so high is that most movies are not informed by a single artist's vision. It is a truism that the greatest films are those where the director was dictator, had unfettered artistic control over the final product. In box-office obsessed Hollywood this has less often been the case than in Europe, which explains the popular belief that European productions are so-called art films, while most American ones are really merely movies. All the great European directors (Bergman, Fellini, Godard, et al) were allowed free rein by their producers. Of course in Hollywood it has happened too†, and continues to happen with greater frequency now that audiences have become more

---

* It's a good joke, but I've recently learned that camels are actually capable of racing at speeds almost equivalent to that of quarter horses. Oh well, substitute llama in the joke if telling it to someone who knows anything about camels.
† The most famous example being Orson Welles, director of the universally acknowledged masterpiece, *Citizen Cane*. A more recent example is the late Stanley Kubrick.

sophisticated, and so those controlling the purse strings are more willing to let a super-star director just do what he wants to do.

.

The crucial point here is one that film critics are very familiar with: A director is more like a CEO than the traditional artist. He does not work alone in his studio or study. He delegates and orchestrates; he is extremely dependent on the competence and talent of others. So credit for the quality of his final product is not his alone—nor is the blame. A poor script writer, a bad cinematographer, an uninspired actor, a weak composer or compiler of the soundtrack—all can ruin a great conception. And a brilliant writer, great cinematographer, fine actors, and an inspired composer can redeem a mediocre film concept.

.

So here we have an art form where the artist (assuming we accept the conventional idea that the director is the one who creates the film) is not necessarily gifted in all—or even in any!—of the component arts comprising the whole work. His creative talent is almost administrative, and unquestionably more critical than creative (at least in the sense of coming up with something new). He selects the script, the actors, the cinematographer, the whole motley crew involved in the project. He delegates the creation of the components of what will comprise the film to these folks, and then decides what from their labours to include and exclude. Later I will address this issue of how much of what is called creativity is actually criticality, even for the individual artist working alone. What is germane here is that for the creation of a film, the dominant multi-media art form of our time, the artist need not be a master of any of the component arts. His creativity is of an entirely different kind: it is in the ability to select and control his support group of artists and to integrate what they create into a meaningful whole.

.

As science has become less an individual project and more a collaborative effort, so too it seems has multi-media art. The film director as artist represents a new and unique conception of what artistic creation entails.* It also points to one solution to the problem initially raised about unreasonable expectations of wide-ranging talents in any artist presumptuous enough to create multi-media art.

.

Opera.

---

* Directors of theatrical productions have a right to take issue with this statement, which of course applies to them as well to only a slightly lesser degree. My apologies.

I never had much interest in opera, even after I developed a taste for concert (so-called 'classical') music, but I'm slowing coming around to appreciate it, largely because of an increased understanding of the historical importance of it in the development of what is called 'absolute' music, the traditional repertory music played at Carnegie Hall or The Philharmonie Berlin—e.g., Mozart and Beethoven and Brahms concerti and symphonies. An understanding of the development and appreciation of music from the Baroque Period into the Twentieth Century is illuminating and useful in understanding the problems associated with the creation of multi-media.

The art form known as 'opera' is a relatively recent invention. Jacopi Peri's *Dafne* is traditionally considered the first opera, and it was written in 1597, at the beginning of the Baroque Period in concert music. Of course, depending on one's flexibility in defining this art form, ancient Greek theatrical performances could be considered opera, as could modern Broadway musicals or even films such *West Side Story*. The audience for early opera was aristocratic, educated and musically sophisticated. The dominant genre of the 17th century was known as *opera seria* and *was* serious, usually dealing with classical themes. But with the rise of the middle class that accompanied The Enlightenment, another genre called *opera buffa* (comic opera) rose to prominence. By the early 18th century, going to the opera was more like going to the movies fifty years ago than like going to the opera now. (Tickets were cheap and there was no implied formal dress code.) I don't think they had popcorn back then, but they certainly had munchies and liquid (usually alcoholic) refreshments. Apparently they were even more boisterous and rowdy than teenagers at a Saturday move matinee in the 1950's. But then at the end of the century Richard Wagner came riding up on his high horse. For Wagner, opera was the ultimate multi-media art form and very, very serious business. It combined all the great arts from literature through visual art to music. Wagner, no modest fellow, and he felt his wide-ranging creative genius was up to the task of producing huge, complex masterpieces in this art form.*

The parallels between multi-media opera and cinema are obvious, albeit an over-simplification. There is opera seria and there is the art

---

* Once again, I must apologize to all the musical historians who are now suffering apoplexy because of this presumptuous, simplistic one paragraph history of opera.

film. There is opera buffa and there is the movie. There is Wagner's *Ring Cycle* and there is Lucas' *Star Wars Saga*.[*]

The libretto for an opera is roughly equivalent to a screenplay, for it includes stage directions and all the textual, non-musical, elements, the dialogue. (In an opera this means the lyrics to the musical numbers, the arias, and the spoken passages, the recitatifs.) There was no expectation of the composer of an opera having the literary talent to write the libretto. Throughout most of the history of opera, the librettist collaborated with the composer—to a greater or lesser extent. And the music got first billing. Name any opera, and of those who recognize the name of it, most will be able to name the composer. Very few, except opera buffs, will be able to name the librettist. Wagner, however, did everything, including write his own librettos. In most movies, the director does not write the screenplay, and here too the script writer—or often writers—rarely is given much credit or notice. But in many of the films that we have come to revere as serious works of art, the director either wrote the screenplay or intervened with the writing to such an extent as to be considered a co-author. Directors such as these, the Bergmans and Kubricks respectively, are the Wagners of cinematic art. But such diversity of talent is rare, and so too are films or operas that so clearly are all of a piece and present a single individual's artistic vision.

Critics of opera often cite the second-rate, often inane, dialogue, lyrics and plots of many operas. Critics of movies do the same. There are plenty of exceptions, of course, in both art forms. (Mozart was fortunate in often collaborating with the brilliant, flamboyant, and eccentric librettist Lorenzo da Ponte.[†]) Furthermore, in opera, the dominant role played by the music can easily redeem unexceptional lyrics.

I have largely avoided the other major issue associated with multi-media art of an artist's creative talent extending to performance of their work. (Most of the great concert music composers began their careers as performers; Mozart only being exceptional in his exceptional virtuosity.) However, to examine this question would be to digress more than I have already.

---

[*] I hear a subterranean rumble as Wagner spins in his grave in response to this comparison.

[†] Da Ponte ended up in New York, first opening a grocery store (which failed), and then teaching Italian Literature at Columbia College. Strange guy.

The central fact to be gleaned from even such a cursory look as the above discussion is that when a creative artist chooses to mix media, he has to either have multiple talents and skills or he has to collaborate. I will be suggesting in the final section of this book, that technological advances, new creative tools, are making the former option more feasible. Scientific advances are compensating for our sensory and our motor limitations to creativity.

BEYOND CULTURAL METAPHOR AND SENSORY LIMITATIONS: MULTI-MEDIA SCIENCE

Creative scientists have their own set of 'multi-media' issues to contend with. Many of the problems are similar to that of artists, such as the need for talent in diverse domains, or, alternatively, having to resort to collaboration with others, to create a unified whole. Another problem is translating their insights into metaphors or images that are comprehensible to us with our limited cognitive and sensory resources.

Consider first the issue of needing to possess diverse creative talents to create a complex, yet coherent scientific theory. Einstein, by his own admission, was not a math whiz. Of course, he was far and away above the average soul, but he was not a great and deep mathematical thinker. Math was a creative skill he needed and add just to the degree required, but it was not central to his creative genius: he was a great physicist, not a great mathematician. Like the film director, whose grand vision requires some creative ability in a great number of domains from cinematography to personal relationships, the scientist too requires a variety of above-average skills in many areas. And as in film, the contemporary scientist more and more has to rely on collaboration with others to piece together his creative vision into a coherent whole.

The second issue is how to translate deep scientific insight into comprehensible metaphors. Scientific explanations are always metaphorical. The metaphors chosen are, naturally enough, those readily available within the current culture. Newton published his *Philosophiae Naturalis Principia Mathematica* in 1687, in which he laid out the basic principles of a strictly deterministic physics, modeled on the complex but inexorable intricacies of clockwork. It is surely no coincidence that public wonder and fascination with it the workings of the first mechanical clocks was in the air when Newton developed his theory. (It was in 1670 that the English clockmaker Willam Clement developed the first very accurate pendulum-driven "grandfather" clock.) To this day the classical physics of Newton is referred to as a "clockwork universe" view of nature.

A contemporary example is the use of the computer as a metaphor for the working of the human brain. Sensory neurons are compared to computer input, the CNS (central nervous system) to the CPU (central processing unit), and the efferent or motor neurons connected to our muscles are compared to the cables connecting our computers to printers or other apparatus that *responds*. Of course, cognitive scientists are not so naïve as to believe our brain works exactly like the circuit boards in our computers, but the metaphor is found useful because computers, like clockworks in Newton's time, are more comprehensible than the actual abstract theories attempting to describe neurological function.

.

As scientists probe deeper and deeper into what makes the universe—and us—*tick*, they are forced more and more into the use of metaphor to describe—and even think about—ideas that are foreign to our sensory experience. No one has actually seen an atom, but a diagram with a ball in the centre with little balls circling it helped us grasp what it is like—at least until quantum theorists wanted to change the diagram to a fuzzy blur. And viewing distant stars and nebulae through a telescope seemed to be very much like looking at something directly, but of course it isn't. And we only got a really good 'look' out into the distant reaches of space with the invention of radar telescopes. Before these devices, we could not 'see' one helluva lot that is both beautiful and interesting. Most people with any aesthetic sense of awe are moved by NASA's dramatic, colourful images of distant nebulae. But, of course, these are so-called—and quite accurately—*false-colour* images; i.e., images created by arbitrarily assigning a wavelength in our visible spectrum to each of the wavelengths the telescope actually detected. It makes it easier to distinguish details and differences and is certainly aesthetically pleasing—but if you took a rocket ship out to take a closer look at one of these nebulae, it wouldn't look at all like those spectacular NASA images. We forget that even something as basic as colour does not exist except in our brains. An apple, you say, is red. Then turn down the lights in the room. Eventually it becomes gray. Red isn't in the apple. It is in our perceptual experience which is dependent on the actual wavelength of the electromagnetic radiation causing, or not, changes in chemicals in our retinal photoreceptors, which eventually change chemical balances in the neurons of the visual cortices of our brains.

.

In mathematics especially (and mathematics is now at the core of our understanding of the physical world) the ideas essential to current understanding are so beyond our sensory and intellectual capabilities that it is only through metaphor (which by definition is a distortion, albeit an enlightening one) that we can even attempt to deal with them. Even something as 'simple' as a four-dimensional object is incomprehensible, yet current theories in physics are postulating eleven-dimensional 'strings' (another metaphor) as the building blocks of the universe. Clever computer programmers build fascinating simulations of hyper-cubes (four dimensional cubes) to try to help us get a handle on the idea of four physical dimensions, but of course in viewing these we always have to be switching between different views within our three-dimensional comprehension. And even there the three dimensions are being simulated, as in perspective painting, on a two dimensional surface: the computer screen.

.

Oh, our brains are such very limited tools for apprehending the world! How to the fill in the picture in all its incredible complexity? The artist may turn to multi-media creations to try to populate the far corners of the canvas. The scientist may turn to artistic and metaphorical media. But in the end both must inevitably fail to apprehend and communicate the wondrous whole of their visions.

.

Nevertheless it is our very deficiencies—as a species and as individuals—that inspire and focus our efforts.

CASE STUDIES: STEPHEN HAWKING, LUDWIG VAN BEETHOVEN

We're not only all brain-damaged in some way, but we're also each deficient in some individual way in terms of the input to—and creative output from—our brains. In short, we're all handicapped—or to use the politically correct euphemism—'challenged'. One of my own many input handicaps is an inability to distinguish phonemes associated with other languages: I do not have an 'ear' for languages—to put it mildly. One of my own output handicaps is clumsiness with tool usage: I always find myself bleeding five minutes into any handyman project.*

Nevertheless, some people's handicaps—or challenges—are sufficiently extreme to elicit pity. (Empathy would be better, but pity is the usual response.) Two such extreme cases are here considered. And in both cases the individuals more than met the challenge.

Science is based on observation, on going out into the world, exploring and experimenting and getting one's hands dirty in the process. In this way science is usually set up in opposition and contrast to the proverbial philosopher plunked down in his armchair trying to think his way through to deep knowledge, without even looking out his window. So then what greater handicap could a scientist have than paralysis? Yet one of the great scientists of our time is doing his science from a wheelchair: Stephen Hawking.

And what greater handicap can one conceive of for a composer than deafness? Yet who was one of the greatest composers of all time?

Both Beethoven and Hawking exemplify in an extreme degree the extent to which human beings are able to transcend creatively not only the universal limitations of human perception, but also exceptionally debilitating limitations imposed on them by disease.

Stephen Hawking was an athletic young man. He was an equestrian and coxswain of a rowing team at Oxford. Then at twenty-one years of age, he developed problems with co-ordination that led to a diagnosis of amyotrophic lateral sclerosis (ALS or Lou Gehrig's

---

* At our mortgage burning project, my friends gave me 'upkeep' gifts. The most apt was a pair of gloves—both for the left hand.

Disease). This is an untreatable disease of the motor neurons. Doctors at the time gave him three years, at the most, to live.

To his good fortune, and humanity's, the doctors were wrong. He is now, at time of writing this, well into his sixties, although the disease has relentlessly progressed, leaving him so paralyzed that his voluntary movements are limited to blinking and slight facial twitches. Yet with the help of computer technology, he is able to talk (through a voice synthesizer), write research papers and carry on email correspondence with other scientists.

Hawking's motor neurons may be damaged, but the interneurons of his brain are in fine shape. He is currently Lucasian Professor of Mathematics at Cambridge, a chair once held by Newton and an honour based on his great contributions to theoretical physics. If someone were asked to name three great physicists, I suspect that Newton, Einstein and Hawking would be the three named by more than ninety percent of those asked—at least those who could actually name three! The fame is not unjustified in terms of his real scientific contributions, but is probably largely because Hawking's name is associated with the weird cosmological phenomenon known as a black hole that has captured the public imagination. He, like Einstein, produced theories that seemed esoteric, counter-intuitive, and then explicated them to the general public with examples that fired the imagination. One immediately thinks of Einstein's "Twin Paradox" where round-trip space travellers age differently than stay-at-homes, and thus travel in time as well as space. When have we not been interested in the paradoxes of time and time travel? With Hawking, it was the idea that "black holes have no hair"*, a catchy phrase to describe those mysterious, invisible, collapsed stars, those cosmological monsters that consume everything forever, including light and information and even time itself—which stands still at something called an 'event horizon'. What could more fire the fearful imagination than the end of the time and the universe?

Both Einstein and Hawking were quite willing to talk in plain language to us common folk about these arcane matters, matters that could only be deeply understood by their colleagues. Einstein wrote several books for the layman, notably *Relativity: The Special and the General Theory*, and Hawking wrote *A Brief History of Time*, which is probably the only physics book to ever make the best-seller list. Of

---

* Hawking has changed his tune regarding this pronouncement, but that is another story for another writer.

course both Einstein's and Hawking's striking public persona was also a factor in their fame.

I have argued that our general and individual human limitations have to be acknowledged in doing both art and science, but also that some of these limitations can be overcome and some even paradoxically result in increasing our creative potential. Stephen Hawking is an example of both.

By all accounts, including his own, Hawking was less than a highly motivated student, more interested in a good time and enjoying his then robust physicality than in his formal studies. While he definitely excelled in his studies at Cambridge, he only did so because he was so brilliant—not because he worked at it. (Fellow students talk about taking home problem sets that they worked on for weeks, solving only a decent percentage of, while Stephen ignored the assignment until the night before solutions were expected—and then just locked himself in his room and worked through all the problems in time for morning class.)

It was only when his disease struck him that his motivation toward intellectual accomplishment increased. It is a terrible irony, like many associated with creative accomplishment, that the very thing that destroyed his chance for a 'normal' life may very well have been the cause for his very exceptional accomplishments. He has virtually said this himself. He also has frequently referred to himself as "lucky", which should give us all pause when feeling sorry for ourselves because of some specific handicap or misfortune.

Of course, in one very specific sense he was 'lucky'. He was doing science at the most extreme theoretical end of the spectrum from dirty-hands experimental science. Hawking, like Einstein, could do "thought experiments" (*Gedankenexperimenten*) that would give us as much insight into the real world as any laboratory or field work. (Hawking originally intended to study mathematics, which is as far from empiricism as one can get.)

And he was also lucky in that the advancements in computer science and technology made it possible for him to live in his mind while retaining contact with the outside world. Anyone who doubts the contribution of science and resultant technology to increasing our creative potential should read a biography of Stephen Hawking.

Ludwig van Beethoven is another story. There was no technology to compensate for the devastating disease that left him deaf. And his ability to work in his mind without the presumably prerequisite sensory auditory input is less comprehensible than Hawking's ability to work in his mind without control of his motor output. But some biographers have suggested that Beethoven's disability, like Hawking's, may have been what motivated him to monumental achievement.

.

Louis, as his friends called him, was born in 1770, twenty years after Bach died and fourteen years after Mozart was born. These dates have relevance. Mozart had epitomized the 'Classical Period' of concert music, just as Bach had epitomized the 'Baroque Period', and whose death in 1750 marked the end of that era. So Beethoven was to epitomize the transition from The Classical Style to The Romantic Style, and this radical transition was to occur at least partially because of the deterioration of his hearing beginning around 1800.

.

Beethoven did not have a good childhood, for his father was by all accounts a nasty, ambitious, alcoholic, and his son Ludwig, sad to say, in some ways was an apple that didn't fall far enough from the tree, for he was less than exemplary from a moral standpoint, which is a gentle way of saying he also often behaved abominably. Those interested in the sordid—albeit sometimes darkly amusing—details need only read any serious biography of this great composer's life, where well documented will be everything from his conning his compassionate mentor "Papa" Haydn to his ruining the life of his nephew Karl because of an inane family quarrel.

.

That Beethoven had serious mental and physical problems before the onset of his deafness is indisputable. He had delusions that he was the illegitimate son of Frederick the Great and changed the 'van' in his name to 'von' because the latter indicates royalty rather than mere Dutch heritage. His physical problems included serious problems with his intestinal tract, and the bizarre fourth movement of his Second Symphony is widely accepted to be comically 'programmatic' or 'onomatopoetic' of his flatulence and other digestive issues! His autopsy indicated that his liver, gall bladder, spleen, and pancreas were all diseased, although actual cause of death remains unknown.

.

Beethoven had started his career, like Mozart, as an accomplished performer, before moving into composition. Unlike Mozart, who seemed to have compositions delivered to his brain fully conceived,

Beethoven had to work at it. It is said that Mozart simply heard a fully formed composition in his mind's ear and then merely had to perform or transcribe it. Enough workbooks of Beethoven are extant to know that he composed in an entirely different way. He would take a single motif and work it through numerous transpositions, inversions and variations along the way to the completed theme. (His Fifth Symphony is often cited as example of this brilliant development of a basically banal motif into a grand theme.) Tunes apparently just popped into Mozart's head, but this was not the case for Beethoven. Nevertheless, he still must have been hearing them in his mind's ear, for by the time of his grand Ninth Symphony he had become almost totally deaf except for the very lowest frequencies.

Musicologists divide Beethoven's compositional life into three periods: "Early", "Middle", and "Late". The Middle Period is usually said to begin with his Third Symphony, the *Eroica*, completed in 1805, and to coincide with the serious onset of his deafness. This symphony is a major departure from the classical style of his first two symphonies, and it marks the beginning of a major revolution in the whole conception of the function of music. Beethoven's so-called "heroic style" of this period transformed concert music from mere entertainment to an expression of deep emotions. Although Beethoven's productions of his Early Period are brilliant, his great reputation and the importance he is accorded in the history of western concert music is based on compositions from this time on. And that this is also the time in his life when he was in profound personal crisis strongly suggests that this was the onset of profound, devastating disability that focused his native creative energy to a laser intensity that blinds us still with its brilliance.

Another crisis was to mark Beethoven's transition into "The Late Period" where he composed the Ninth Symphony, the late string quartets, his last five piano sonatas, and the *Missa Solemnis*. His popularity had declined as had the number of his patrons around 1815. He began behaving even more erratically and outrageously, insulting friends and supporters, and setting out on the warpath regarding custody of his nephew Karl. He could no longer conduct, never mind perform, for he couldn't even hear clearly what the orchestra was doing. But again this crisis seemed to kindle a creative rebirth.

The works of this period are extremely radical and innovative, and once and for all replaced the rules and conventions of the classical style with the idea that anything goes, for music is to be the expression of deep emotion and ideas, not mere formulaic, albeit clever, entertainment. The crowning achievement of this period, and one of the most influential concert works ever written, was his Ninth (or Choral) Symphony, composed a whole decade after the almost conventional Eight.

.

It is hard to imagine Beethoven, like Hawking, saying he was "lucky". But more than one musicologist has suggested that *we*, at least, were lucky: that it was Beethoven's aural isolation that made possible his greatest works. Once again we have the depressing paradox that, at least sometimes, creative persons' troubles are a contributing factor to their accomplishments.

MEMORY, LANGUAGE, AND THOUGHT: OUR UNMETRICAL MENTAL MAPS

"The charm, one might say the genius of memory, is that it is choosy, chancy, and temperamental: it rejects the edifying cathedral and indelibly photographs the small boy outside, chewing a hunk of melon in the dust."
—Elizabeth Bowen

.

"If ever asked what one 'has in mind', the only right reply is 'More than I can say!'"
—Hippokrites

We make what we make from what we have at hand—or, more accurately—what we have in mind. In many universities, psychology departments are being replaced with "cognitive science" departments. After psychologists finally acknowledged the limitations of behaviourist theory, with its stubborn refusal to even consider the study of cognition as scientific, the pendulum swung back. We *are* thinking creatures. Aristotle, that premier biologist, had it right when he observed that each creature on the planet had specialized skills: the bee's navigational skills function for the harvesting of nectar. Function, he argued in his *Nicomachean Ethics,* is a characteristic of all things, animate and inanimate, and "the function of man is a certain form of life… the exercise of the soul's faculties and activities in association with rational principle" In short, what defines us as human beings is *reason*, a specialization that has been refined to the greatest extent in *Homo sapiens,* and the function of reason is the good life. We think, therefore thinking (and creating) are our functions in the grand scheme of things. And even if there is no 'grand scheme", thinking or cognition certainly are deserving of scientific investigation, as well as artistic study.

.

Just as with perception, cognition is largely a product of filtration—or self-censorship, if you like. Even that infinitesimal fraction of events in the 'real world' that we detect—and which somehow make it all the way to our brains—even these tiny fragments are then subject to further distortion and deletion.

THE CASE FOR FORGETTING: ROLE OF MEMORY IN THE CREATIVE METHOD

So much depends on memory! But then, perhaps not so much our identity (as once supposed), as our creativity. And, perhaps not so much of our creativity is grounded in memory (as once supposed) as it is in selective loss and distortion of memories.

.

We think by manipulating the components of our memories, but what exactly is memory? Current theories of memory use the commonsensical, tripartite paradigm of input, storage, and retrieval. Failures of memory can have their source in any of these stages. Information may be lost in the transition from input to storage. Information can fade from storage. Or information can remain in storage but not be available for retrieval.

.

Consider the input stage. Again this is considered a tripartite process. The first stage is sensory memory: the raw data from our senses hangs around for a very short time: visual input for less than a quarter of a second (virtually the blink of an eye), and auditory (or 'echoic') for, at best, two seconds. Some very small portion of this data will be moved to short term—or 'working'—memory. How long it stays there depends on how intensely we attend to it. For example, if someone says a phone number, it will only go into working memory if we attend to it, and only stay there if we 'rehearse' it; i.e., say it over and over to ourselves. Should our attention be distracted, it 'becomes history'—unrecorded history. This fragility of our working memories is disturbing, but equally disquieting is the fact that it is a very small storage space. There is a classic research paper by George Miller entitled "The Magical Number Seven, Plus or Minus Two: Some Limits on Our Capacity for Processing Information" in which he presents solid empirical evidence that our working memory can only hold nine, at best, but often only five, discrete items at the same time. If we don't quickly clear our mental desktop, new stuff dumped there is going to push older stuff off onto the floor, where it will be promptly swept up and discarded.

.

Long-term memory consists of the stuff we file away before it gets swept away. (One of the most important brain structures responsible for this filing is the hippocampus.) But what this long-term storage facility is like, how it is organized, to what extent the contents of its rooms decay and turn to dust—these are as yet unanswered questions still being investigated.

.

The third component of memory is retrieval from our long-term storage facility. Memory sometimes fails us. Why? Is it because the stuff eventually got thrown out because it wasn't being used? Or is it because we just can't find it amidst all the clutter? What is sometimes called "The Junk Box Theory Of Memory" maintains that it is usually the latter that is the reason we can't remember something. Supporting evidence for this theory is our ability to eventually, sometimes inappropriately, retrieve something after we need it. I'm having a discussion with my wife and friends and trying to remember the name of the actress in a particular movie, but I just can't seem to do so. Then in the middle of the night, I wake up and cry out "Julia, Julia Roberts!"—and my startled wife gives me a whack on the side of my head. Obviously, I must have had Julia stashed away up there in some back room of my memory, but I just couldn't find her when I needed her.

.

Now undoubtedly some memories do disappear entirely. Brain cells die and associated neural connections are lost. But we tend to overestimate how much is really gone, as opposed to it being there but merely hard to find. This is shown in studies of language learning, or rather 're-learning'. You learn Italian in your home town of Milan, and then at a very early age your parents immigrate to Toronto where you are educated and become totally assimilated into North American English culture. Your highly educated parents, who are bilingual, switch to English as the language spoken at home as well. Then in university, you develop an interest in the rich literature of your homeland, but you can't remember more than ten words of Italian, so you sign up for an introductory course in Italian. You're normally just an average student, but you quickly move to the head of the class. You 'relearn' your native tongue at a rate that clearly suggests those neural connections that constituted your memory of Italian have not disappeared, even though it certainly felt like they had.

.

There is another misconception about memory that is important. Just as we often mistakenly think we've lost a memory forever, we simultaneously place unwarranted confidence in the accuracy of the memories we do find. This is especially true of 'episodic' memories. Episodic memories are memories of life events and experiences. They are often contrasted with 'semantic memories', which are memories

100

of facts and information.* This dichotomy is, unlike too many psychological distinctions, one really based on solid neurological evidence. Distinct parts of the brain are clearly responsible for these two types of memory, as the strange phenomenon of retroactive amnesia demonstrates. Some mugger whacks you on the head, and you wake up unable to remember your past. But if in your past you were a chemist, you'd still be able to recite the Periodic Table of Elements flawlessly. You just wouldn't remember where and how you learned it. The brain trauma affected the part of your brain responsible for episodic memory, but not semantic memory.

.

Our episodic memories are saturated with emotional colour, while semantic memories are not. When we remember past events in our lives, we remember the associated feelings. When we remember facts, they are colourless. Dinner was late last night, and you remember that, but you also remember feeling annoyed. When you recall the second element in the Periodic Table, it has no emotion attached to it.† Most people are far more willing to admit they aren't sure about a remembered fact than a remembered event in their life.

.

What most people don't realize is that memories are not static: they are dynamic. An analogy is useful here. Think of remembering something as the act of going to your mental filing cabinet and retrieving a document. Maybe the document is a bit faded and hard to read, or maybe in reading it you are a bit careless, but in either case often you inadvertently *edit* the document. Then this revised version is re-filed. One disturbing implication of this is that the more often you remember, retrieve, a memory, the more likely it is that it has been changed. Someone once remarked that the purest, most accurate memories are those locked away in the amnesiac's brain, for they have not been taken out and messed with.

.

This editing process is more likely to occur with the emotionally coloured paper of life events, for unlike hard facts stored in semantic memory, such documents are not likely going to be confirmed and

---

* Both episodic and semantic memory are subdivisions in what is called *declarative* or *explicit* memory. The other major division is *implicit* memory which includes memories which are not conscious, such as learned sensori-motor skills (how to ride a bike) or conditioned responses (Pavlov's dogs drooling when he rang the dinner bell). These implicit memories are not particularly relevant to the role of memory in creativity.

† Unless for some reason helium excites you! (Perhaps because as a child had so much fun with balloons that when your mother told you they were "full of helium", you came to attach the emotion of glee to the word 'helium'.)

clarified by comparison with present reality. If my retrieved document is about the second element in the Periodic Table and I can only clearly read the first letter 'H' and I say "Hydrogen", I'm likely to have my misreading corrected by my chemistry teacher. However, if my retrieved document is about an event in my past, it is highly unlikely that any misreading will be corrected before I re-file the edited document.

.

This was brought home to me by an embarrassing discovery I made about a vivid personal memory. My son was born by Caesarean Section, the necessity of which was only confirmed when my wife's obstetrician showed up at the hospital many, many hours after she had been admitted for what he'd diagnosed over the phone as probably 'false labour'. This 'false labour' turned out to be very real labour and caused her incredible pain and suffering since my son hadn't read the manual explaining how to exit a womb and was a 'brow baby' stuck in the birth canal. I 'remember' most vividly when the doctor finally arrived, all decked out in tacky golfing gear, including garish shorts. I also remember being extremely angry with him. My wife was in no condition to really observe and remember anything, but she over the years had confirmed me in my memory of this long, important night in our life. It was only many years later while celebrating my son's thirteenth birthday and talking about the night of his birth that a light came on in my head. Something was seriously wrong with this extremely vivid 'memory'. My son was born in Toronto on January 25th (in the dead of a Canadian winter) at around midnight (in the dead of night). So where the hell could our doctor possibly have been playing golf? At an all night, indoor mini-putt for obstetricians on-call?

.

When Freud was still in short pants, the home-spun American humourist Josh Billings was showing more insight into memory than Sigmund ever would: "There are lots of people who mistake their imagination for their memory." Freud gave us the very dubious and dangerous idea of 'repressed memories'[*], while Billings with this offhand remark gives us the very real insight that memories aren't repressed; they are, on the contrary, often invented. And if not invented from whole cloth, at least so edited and revised as to bear virtually no resemblance to the original experience.

.

---

[*] The harm to innocent people caused by clinicians guiding children and even adults into inventing 'repressed memories' of abuse is finally being documented.

Somewhat paradoxically, this serves the creative person well. My wife insists that every time I retell one of our modest adventures, it gets more outrageous and exaggerated. I prefer to think it just gets better—far better than the original experience—but a part of me also stubbornly refuses to believe my latest version isn't the god's honest accurate truth of what happened.

.

Artists and scientists need experience from which to build what they create, but paradoxically a perfect memory of these experiences is among the worst things with which a creative person could be cursed. There are people with unbelievably good and accurate memories, and they are without exception entirely *uncreative*, and, far more often than not, also dysfunctional in every other way. Consider the man on whom the movie *Rain Man* was based. The real life Rain Man (Kim Peek), like most memory *savants*, is autistic. His well-documented prodigious memory includes the ability to read a book in about an hour and really recall all the details from the text when objectively tested.* Allegedly he can remember, virtually verbatim, content from over 10,000 books. Those of us teachers who occasionally find ourselves in front of class trying desperately to remember a simple name or date or quotation—that we will presumptively expect our students to recall on an exam—might understandably be envious of such a talent. But Peek has difficulty understanding the simplest metaphor, and certainly could never integrate his huge mental library of knowledge into even the simplest exposition on any of the topics he undoubtedly 'knows' more about than anyone else alive.

.

When Einstein was asked, as an aside in the midst of a mathematical question about his General Theory of Relativity, what was the speed of light, he replied that he "didn't know" and that when he really needed that information for his calculations he "looked it up." All right, the story is probably apocryphal, or he was joking, but either way it makes a valid point. Just as the greatest creative minds likely would score less than spectacularly high on conventional IQ tests, so too they may perform no more than average on tests of memory, even memory involving the very stuff with which they create their art. Wordsworth's famous poem "Daffodils" reveals on close analysis what seems like an exceptional skill at natural observation and memory, but it was his sister (who accompanied him on the walk that inspired the poem) who had the sharp eye and good memory to

---

* Unlike Woody Allen who failed to show such detailed comprehension at high speed reading: "I took a speed-reading course and read *War and Peace* in twenty minutes. It involves Russia."

supply him with the material from which he shaped the poem. Poets and writers are often credited with being great observers and recorders of the teeming life around them, but this is probably giving credit where credit is *not* due. Artists repeatedly admit that it isn't what you put in, but what you leave out, that really matters. This applies to memory as well. The best strategy, which seems natural to the artist, is to *not* notice and remember most things, but to etch into memory and then elaborate what is most salient or useful. The portraitist, who can capture in just a few pencil strokes on his sketch pad the essence of a person he met yesterday, may very well have this ability precisely because of his ability to forget irrelevant, undefining details of the face. To formulate his theory of relativity Einstein had to 'forget' about the alleged distinction between a person in a free-falling elevator and a person floating in outer-space orbit.

Finally, it should be pointed out that John Locke's hypothesis that memory is what defines us as distinct, unique individuals is demonstrably wrong. Sufferers from retrograde amnesia who have lost all memory of their past still retain their pre-trauma personality traits. Memories are dynamic, ever changing, ever undergoing revision, elaboration and deletions and so are our 'selves'. Even our bodies (including our brains) no longer contain more than a few molecules that were present there when we were seven years of age. As Steven Pinker points out in his fine book on language acquisition, *The Language Instinct,* our genes and experiences don't just store up stuff that finally accumulates to *define* us. Rather, what is useful is retained, and what isn't—well, it is recycled, rebuilt into something new, something that may have current usefulness, or else it is simply discarded to leave room for something new in our cognitive workspace. In this sense, the existentialist philosophers certainly had it right: existence precedes essence. We do indeed create ourselves anew every day. If nothing were discarded or revised, nothing new could be created. Our minds and lives would simply be a meaningless accumulation of memories—an ever-expanding junkyard.

A LEGEND FOR OUR MENTAL MAPS: LANGUAGE, COGNITION AND CREATION

I speak, therefore I think. And if I think, therefore I am. Ergo, it is speaking, *language*, that is the prerequisite to existence as a rational being. There is no idea of 'Mother' in Baby Jane until she utters 'Mama!' Or so goes the argument now to be considered.

.

One of the important philosophical debates, dating—but of course—back to the ancient Greeks, is about the role language plays in cognition. At one extreme are those who claim that thought is impossible without language, while at the other extreme are those that claim thought is what creates language (as a useful tool in some circumstances), but maintain that thought can and does constantly occur in the absence of language.

.

Once again it is necessary to be clear in defining terms before taking sides in this debate. In the spring, male birds sing a particular love song, and in doing so communicate their lust and availability. This is undoubtedly communication, so if one defines language simply as communication, then the horny male chickadee welcoming spring by singing "Hey sweetie! Hey sweetie!" has language.

.

"Music is the universal language!" is a common cliché. If so, then I'd like someone to compose a piece of instrumental music explaining how to put together the unassembled cabinet I just bought and came, not with a music CD, but with written instructions for assembling it. (in five languages).

.

And then there is so-called "body language" and gestures, some of which are universal among humans and some not. Blind babies still smile and frown to communicate their emotions of pleasure or displeasure respectively. But that North American thumbs-up gesture indicating a positive affirmative reaction to something (or the hitchhiker's desire to get a free ride) is interpreted as the equivalent of the 'giving the finger' in Greece. 'Thumbing' a ride in Greece is more likely to get you run over than picked up. Until recently, most Bulgarians indicated agreement by shaking their head, and nodded to indicate disagreement.*

---

* I know this from personal experience while traveling in Bulgaria. My wife's poor German managed to convey to a local, with a similar smattering of German, our question of whether we were going in the right direction to the hotel. But his "Ya!" reply was accompanied by vigorous shaking of his head. We trusted to his verbal—

Those who study language make important distinctions between phonology (a language's sound system of phonemes); morphology (the rules for word formation, such as adding an 's' for plurals in English); syntax (the rules for the ways words are combined to form meaningful sentences); and semantics (the meaning of the words and sentences). At what stage do we admit to true language being present? (And according to some, only then admit to true cognition?)

My dogs can recognize words and even understand simple sentences with different direct objects. "Go get Wilson!" sends my dog, Maggie, off after her football, while "Go get Lion!" sends her after her stuffed lion toy. Was Maggie unable to think about her football or her stuffed toy before she was taught the human vocalization associated with them? Does she have 'language'?

Researchers investigating primate language maintain that it is the location of the voice box, rather than any cortical deficiency which is the primary explanation for lack of language in the higher apes.* The great apes possess areas of the left temporal lobe equivalent to those areas in humans that are devoted to understanding and creating speech.† So the work-around to this physical disability has been to teach chimps sign language—or, of all things, computer touch-screen entry! It works, at least to some degree. Even those researchers who set the definitional bar for language quite high, maintain that their primate students, once given these aids, can learn to use language at a level equivalent to that of a young child. The chimp, Washoe, raised with a human family and taught American Sign Language, is the early poster child for this research. Allegedly he could not only distinguish significant syntactical distinctions as that between "Washoe tickle you!" and "You tickle Washoe!"—but also construct them! Another chimp student, Miss Lucy, demonstrated her linguistic ability to construct new adjectival phrases to describe things for which she had not been taught any words: On first tasting a radish, she spontaneously labelled it "cry hurt fruit." More recently a pygmy chimp (a bonobo ape) named Kanzi has replaced Washoe as the

---

rather than gestural—response, which turned out was what he intended and got us to the hotel.
* Like human babies up until the age of three or four months, primates' voice boxes are located in a position that doesn't allow for the sufficiently complex sound production required to create all the distinct phonemes used in speech. Some evolutionary theorists credit the change in location of the larynx as one of the most critical developmentd in human evolution.
† Broca's area and Wernicke's.

linguistic genius of the primate world. Allegedly Kanzi has learned 348 lexigrams, recognizes over 3000 spoken English words, and routinely constructs syntactically complex sentences to communicate with his trainers.[*]

.

There are well informed sceptics about all the aforementioned claims for primate language. Most notable, and very convincing, is Stephen Pinker, who discusses this research in his book *The Language Instinct* and concludes the researchers are doing a lot of wishful thinking and projecting. But even if we do swallow the primate language researchers' claims (perhaps with a large grain of salt), it doesn't affect the argument. Who would claim that Kanzi's alleged acquisition of language (at a level that only those with the most narrow definition of language would disallow as true language) transformed him from an unthinking animal to a creature with cognition? That is simply too absurd. But, it is not absurd to suggest that whatever acquisition of language he really has achieved did *help* him 'think' more effectively, and so manipulate his world. He can now ask and receive (from his loving trainers) what he thinks about wanting, just as my dogs can facilitate getting their thirst quenched by using the 'sign language' of tipping over their empty water bowl.

.

Language in whatever guise (be it gesture, symbol, speech, metaphor, or mental image) simply serves as a tool used in thinking. Language is only prerequisite to thought in the same way that a specific tool may be required for a very particular task. More often than not, should we lack the tool, we find a substitute. If I need to nail together some contraption I've thought about building, a hammer is certainly going to make the task easier, but I could blunder along and complete the task using some substitute available to me—such as my wife's treasured cast-iron skillet. The tool used may be so inadequate as to result in something as poorly constructed as most of my carpentry projects, but I could still proceed with the project. On the other hand, without the tool of calculus, I cannot solve a differential equation. Without the tool of literacy, a "mute inglorious Milton" can never write an epic poem.

.

Lack of language tools is undoubtedly a handicap, just as a well stocked toolbox is unquestionably an advantage. But (as the popular saying goes) "it is a poor workman who blames his tools"—if for no

---

[*] At least according to a report in *The Smithsonian Magazine*, November 2006.

other reason than he should've gone out and acquired the required tools.

·

One person who was convinced that language was essential and prerequisite to cognition was Benjamin Lee Whorf, whose "Sapir-Whorf Hypothesis" was widely accepted by the psychological community for an embarrassing length of time. His claim was that language completely determined thinking. Cultures with few words for colour would not be able to think about the subtleties of colour. Eskimos' ability to deal with the virtually ever-present frozen precipitation of their part of the world resulted in them having hundreds of words for different kinds of the white stuff, and it was this that made it possible for them to construct igloos with the right type of snow.

·

Sound plausible? Well, in 1973, an intrepid researcher named Rosch reported testing the Dani tribe of Papua New Guinea for their skills with colour discrimination. They did just as well as any English speaking person. English contains more words for different colours than any other language, yet the Dani language only has words for black and white. And the Inuit having hundreds of words for snow is an 'urban' myth. In fact, they have no more words than we urban Anglophones who couldn't find the right snow to build an igloo if our lives depended on it. And Whorf's famous claims that the Hopi Indians lack a conception of linear time because their language doesn't address past, present and future in the way English does has been shown to be utter nonsense.

·

Many languages do not have a word for "art", but all cultures have art. And most of the greatest love poems ever written do not use the word "love".

·

Rejecting the Whorfian Hypothesis is not to say that language is irrelevant to creation or that it doesn't influence perception *and* cognition *and* creation. I've already recounted how more profoundly I experienced the landscape of Algonquin Park after having learned to identify and name the trees. The naming of parts is important. A well-stocked language toolkit is going to make seeing, thinking, and creating easier. Certainly Shakespeare would not have been a dolt had his vocabulary been smaller, but there is no doubt his works would've been less rich and resonant. And, of course, he was not afraid to invent words when the English vocabulary didn't have them.

·

This section is about our all-too-human limitations. When it comes to thinking and creating, the bottom line is that thinking surely does precede language and can exist without it, but language is what makes possible full utilization of our cognitive powers—and, of course, the communication of the results of our thinking. What we can build, what we can create, what we can then share, is limited by the language tools we have available with which to work. My dogs can think about being thirsty, but they have to learn to communicate (by tipping over the water bowl) to communicate their thoughts and quench their thirst.

.

Those who have invented new languages, such as Newton's calculus or Feynman's diagrams for subatomic interactions or surrealist representation or musical notation or AI programming languages, are undoubtedly opening our doors of perception, but most importantly they are opening our doors of *expression*.

PRE-COGNITION: MERCATOR PROJECTIONS, ZEN KOANS, AND VISUAL ANALOGIES

The preceding discussion leads inevitably to one of the most intractable questions in all of philosophy: How can abstraction really exist? What is an idea, a concept? And how is it that we are capable of forming ideas? How is it that we can create a category, such as chair or dog or good or beautiful, that allows us to instantly label individual instances as either belonging to that category or not?

Consider the idea of 'dog'. Now biologists have worked out a set of rules for distinguishing and categorizing species. The rules, it has to be said, aren't perfect, for they include such things as inability to cross-breed or dissimilarity in morphology, which aren't always true, but they are pretty good. One could, on the basis of this, argue that categories, ideas, are arbitrarily defined by a set of rules we develop to make the world cognitively manageable. But then explain to me how my dogs, Maggie and Nickel, can clearly distinguish any *Canis lupus familiaris* from a cat or raccoon or squirrel or human child. Anyone who has a dog knows that dogs recognize other dogs as dogs—even though a Chihuahua bears more resemblance to a rat than to a Great Dane and a Great Dane looks more like a pony than, say, the stereotypical dog—a Labrador Retriever. My dogs Nick and Maggie (who are Labs) haven't (to my knowledge) ever taken a course in zoological taxonomy. How can they (who, according to the most generous estimates, possess the reasoning power of a two-year old child) form the idea, the concept, the category, of *dog*? How, for that matter, can a two-year old child form that concept?

Plato, of course, had his dubious Ideal World hypothesis. And current theorists in cognitive science have their theories. The famous Swiss developmental psychologist, Piaget, explains concept formation in young children in terms of 'accommodation' and 'assimilation'. Little Johnnie, starting in on developing a vocabulary, calls every animal he encounters a dog because his mother refers to their pet German Shepherd, Duke, as a dog. But then his mother tells him their cat, Whiskers, is not a dog. The racoon out in the garbage is not a dog. Only Duke is a dog. So then another dog comes to visit, a toy poodle, and his mother calls that creature a dog. Hmm. So he *assimilates* this small, curly haired creature into his idea of dog, and *accommodates* the characteristics of this un-Duke like creature into his idea of dog.

Most cognitive psychologists tend to view concept formation as "prototype matching". This means we form what Piaget would call a 'schema', a prototype of some idea, against which we then empirically test each new case we encounter to see how well it matches this prototype.

However, it is hard to believe that my two dogs did any kind of assimilation or accommodation or prototype matching to learn the idea of dog—as hard to believe as that they slipped out at night to take a course on zoological taxonomy. From when they were pups they damn well knew another dog of any description when they met one.

I'm not suggesting that the studies of cognitive scientists and philosophers on concept formation are misguided or wrong-headed. I *am* suggesting the possibility that, like with language acquisition, we may be hard-wired to form concepts and so can do it without the higher intellectual capabilities that come with age. Even children or animals with a mental age of two, while totally unequipped with the logical tools for learning grammar, or that of defining categories and concepts, still can do both because of this innate and useful pre-programming.

The creative individual may be exceptionally able to tap into these innate circuits and thus form new ideas 'instinctively', rather than through such plodding logical processes as prototype matching. "A new idea springs to mind." This metaphor rings more true than anything like "I came to this new idea by careful deductive and inductive reasoning."

So many really good and radical ideas are pre-intellectual, by which I mean outside the realm of analytical, critical thought—an ability that we only acquire (and alas only some of us) in our teens. This is not to say mature analytical reasoning isn't essential in creativity; it is only to say that the origins of real creative insight seem to come from somewhere else, some place outside of conventional rational thought.

Zen koans are an excellent example of insights that defy—or rather exist outside of—rational thought. What is the sound of one hand clapping? Hard-headed philosophical positivists such as Ayers or Wittgenstein would dismiss this Zen 'riddle' as meaningless. But why then does it resonate so? Why can it be used to induce meditation about such deep questions as the nature of One and Other?

.

Or consider the insights that have made it possible to glimpse and at least partially comprehend and apprehend dimensions that cannot really be presented in their full depth. Our three-dimensional globe is presented on the two-dimensional surface of an atlas, often with what seems the least distortion by the Mercator projection. A four-dimensional hyper-cube can be 'looked at' on a two-dimensional computer monitor display with software that repeatedly shifts the viewing perspective.

.

What is metaphor in poetry (or science) but faulty reasoning? "My luv is like a red, red rose." Come again? She is an animal, not a plant. She, unless sunburned, isn't red. Gravity is the tendency of things to roll down a dip in a rubber sheet made by a heavy object. C'mon. A fundamental force is a geometrical distortion in a manifold? And why would anything roll down into the dip unless something was pulling it there? Gimme a break! Not logical!

.

We use visual analogies to express what can't be visualized. We use words to express emotions and ideas for which there are no words. We use sounds progressing through time, which we call music, to express everything from unnameable abstract aural relationships to emotions.

.

We only get at the deepest understanding and apprehension *and* appreciation of anything through inevitably flawed analogies, synesthesiac comparisons, and metaphors. They are strictly illogical, certainly not precisely accurate by any yardstick of conventional logic. But, nonetheless, they tap into some pre-logical part of our brains and allow us to understand the world and ourselves and at a deeper level than could ever be plumbed by our strictly analytical and logical tools.

.

Just as a two-year-old toddler (who wouldn't know a gerund from a gerbil) intuitively understands grammar at a deeper level than any linguist, just as a dog can recognize another dog without understanding the concept of species, so too do all those who create art and science somehow tap into this pre-cognitive realm for their inspiration—and so illuminate our world. A poem or a piece of music or a painting can 'describe' (and invoke) emotions that are ineffable, otherwise impossible to describe.

THE TRAPS OF INTROSPECTION

Lest what I have proposed in the previous sections be wrongly interpreted as a defence of irrationality, I should emphasize that I consider human rationality, the discovery and application of the rigours and rules of deductive and inductive logic, as one of mankind's distinctive, defining characteristics and accomplishments—along with creativity.

.

However, the final cognitive limitation that has to be considered is our inability to observe and understand even our own rationality or our own mental processes. We cannot see into our own heads as we think and create. It is bad enough that we can never get inside another person's head, but we can't even get inside our own and observe what is going on.

.

An early and disastrously unsuccessful attempt to do so was the method of introspection used by those involved in the early psychological school of so-called Structuralism. Introspection was an attempt to understand cognition by watching ourselves think. Try it. Have someone give you a 'brain-teaser' and try to observe your thought processes as you work on the problem. Or even just try to record what is going through your mind as you read this or daydream or contemplate the beauty of a sunset.

.

Ironically, the competing school of Functionalism, associated with William James, developed the idea of 'stream of consciousness', which was more successful (at least in the hands of writers such as his brother Henry or that other great novelist James Joyce) at capturing, if not classifying and structuring, what is happening in our heads whenever we are conscious.

.

In some ways the problem is similar to that described by the Heisenberg Uncertainty Principle: The act of observing destroys the possibility of observing all the relevant components of what is observed. Probably *complete* objectivity is as unattainable a goal in the realm of thought and creativity as knowing *both* the position and the velocity of a sub-atomic particle. When we watch ourselves thinking we become 'self-conscious' and so are not really observing ourselves thinking, but rather observing our 'self' observing our 'self' trying to observe our 'self' thinking. One's head spins.

.

So the ultimate all too human limitation is our inability to step outside our selves and objectively observe and evaluate what is most important to us: our own cognitive and creative being. But be that as it may, it is also in our nature to try. (Which is why this book.)

THE PITFALLS OF ASSUMED OBJECTIVITY

True objectivity is, of course, a desirable state, but many misconceptions about the nature of scientific creativity have their root in confused views of the nature of the scientific endeavour and its 'objectivity' or lack thereof. Consideration of this takes one into the fascinating but complicated domain of the history and philosophy of science. To understand what creative scientists do, one has to consider what *they* now think—and what they have thought—about exactly what they do, as well as what others think they are doing. At the very start of this book, I suggested that scientific knowledge is simply a different type of knowledge than artistic or sensate knowledge. While I don't think this is a particularly controversial claim, many people would object if I went further and claimed scientific 'information' was just as 'subjective' and personal as artistic or sensate knowledge. Yet, unfortunately, some would applaud such a statement.

At one extreme, are those who view science as a virtually mechanical procedure that consists of the systematic collection of observations of the real world, data from which one logically deduces a broad and objective description of Reality—and so empowers one to predict and control the world. This view of the scientific project can be traced back to Francis Bacon's Enlightenment essay *Novum Organum* (published back in 1620), usually considered the seminal work on the scientific method. Bacon was not a scientist, for even the word 'scientist' wasn't to be invented until William Whewell coined it in the early nineteenth century. Rather, Bacon was a philosopher, and not even really one who could be considered to be a specialist in "Natural Philosophy"—the term used for those precursors to the modern scientist. He was an epistemologist and his conception of the correct path to knowledge was to proceed from accumulation of observable facts to the induction of general laws that objectively defined the nature of reality.* Science was a collective endeavour, virtually mechanical, of collecting data—data from which the general principles describing 'Reality' would eventually leap out. Individual genius or intuition played no part in this process. AI researchers have in fact developed a computer program that allegedly does just this— and they've named it BACON.

---

* He is also well known for his famous and accurate labeling of the four "Idols" of commonplace sloppy thinking: Idols of the Tribe, Den, Marketplace, and Theatre.

However, the historical facts fail to support this view of the nature of scientific discovery. When Kepler was pondering his laboriously collected data on the movement of the planets, there was nothing at all obvious in all that confusing data to suggest that the ellipse was the key to explaining planetary orbits. The insights behind most of the great leaps forward in the history of scientific knowledge are not obvious generalizations that just arose from there being sufficient data. In fact, Einstein's General Relativity Theory had no data whatsoever to support it, and its first empirical confirmation only came half a decade after its publication.

The deep insights into the nature of 'reality' of scientists are just as mysterious and unmechanical and unpredictable as those of artists. They are also just as subject to replacement by new insights. Newton's Gravitational Theory was replaced by Einstein's General Relativity Theory. Is this the same as what happened in the visual arts when Monet's Impressionism replaced Millet's Realism? And even more importantly, does this mean science is as subjective as art or our personal sensate experiences?

Another extreme position regarding the nature of science is that of the social constructionists, with the real fringe being represented by the so-called Deconstructionists, such as Derrida. In what has been labelled the "Science Wars" of the late Twentieth Century, there were numerous, often really ridiculous attempts to claim that science was 'socially constructed' and its 'findings' were mere reflections of the current social and political and intellectual biases of scientists. One deconstructionist even wrote a book claiming that the elementary particles known as quarks were socially constructed concepts—not anything real.* This extreme relativism may seem too absurd to even mention, but it had—and to a much lesser extent still has—currency in certain highly politicized departments of Academe.

The grain of truth to be found deep within the chaff of this once academically fashionable pseudo-intellectual silliness is that scientists themselves have acrimoniously disagreed about whether or not they are discovering objective Truth (note the capital 'T'). Any historian of the philosophy of science can put together two impressive, heavy-hitting opposing teams to battle out this issue. The current official

---

* *Constructing Quarks: A Sociological History of Particle Physics* (1999) by Andrew Pickering.

Copenhagen Interpretation of modern physics clearly repudiates any claims to describing Reality (note the capital 'R".)

The bugaboo of philosophers of science has always been The Fallacy of Affirming The Consequent, which points out that an accurate prediction does not necessarily mean that the assumptions on which it was based are true. My theory is that a Martian has implanted in me an invisible device that can detect when my blood sugar drops below a certain level and that then grabs control of my brain and makes me go out in search of sustenance. Well, lo and behold, it is true that when my blood sugar drops below a certain level, I do get up and raid the refrigerator. Therefore this Martian exists?! Unfortunately, all of science seems to be based on similar questionable reasoning. The indisputable success of science at prediction is unthinkingly taken as evidence that its premises must be correct. In fact, successful prediction is probably the most important criterion applied in judging a scientific hypothesis—and the dirty secret is that it is a logical fallacy! David Hume pointed this out hundreds of years ago in his analysis of causality.

Ultimately the scientific method is, no matter how precisely or imprecisely one defines it, indefensible as a way of accessing ultimate Reality, at least from a strictly epistemological view. Most working scientists only occasionally revert to their original role as Natural Philosopher, albeit the most leading edge scientists are the ones most likely to do so. Most of the time, most scientists simply go about their daily business without giving second thought to such philosophical questions.

But what matters in terms of understanding scientific creativity is not the admittedly important epistemological question of how close to Ultimate Reality the scientific endeavour can take us. Rather it is how working scientists actually work, and if these philosophical concerns affect their work.

The professionalization of science has resulted in the majority of scientists assuming the role of data collectors, as in the Baconian model. "Dial-twiddlers" is the pejorative term applied to those who simply toil in fields of data collection—the majority of professional scientists. They certainly are not philosophers of natural science, nor do they care one whit about such matters. The average experimental elementary particle physicist will pay lip service to the philosophically humble Copenhagen Interpretation, but I doubt very much he doubts the reality of those particles he flings around every day in his expensive cyclotron. Science needs data, and he is doing the important work of collecting it.

However, the theoreticians of science, the ones whose sudden, apparently inexplicable, insights into new relationships change our conception of the world—*they* may wonder whether they are really discovering or merely inventing. And some of them may even see that the easy assumption of their objectivity is dubious. But this is topic for consideration in a later section of this book. The important point here is that the assumption of objectivity is yet another all too human limitation or temptation—and those scientists that somehow try to transcend it, or at least challenge it, are being creative in much the same way as those artists who try to transcend the assumed subjectivity of art by striving for objectivity.

CASE STUDIES: MARCEL PROUST, GEORG CANTOR

Marcel Proust and Georg Cantor are two creative explorers of the outer (or inner) limits of our mental potential. The novelist Proust went beyond where any writer before him had gone in exploring the nature of memory. The mathematician Cantor ventured into abstract cognitive realms that many would warn are potentially toxic to one's mental health. Certainly both were troubled souls. Proust ended his life in self-imposed isolation in a cork-lined room, popping pills and 'living in the past'. Cantor, ex post facto diagnosed as suffering from bipolar disorder, suffered from severe depression in his last years, dying in poverty and despair in a sanatorium at the end of World War I.

Marcel Proust was born in Paris in 1871, son of an affluent family, and was educated at the Lycée Condorcet. Initially he studied law, but soon gave that up to simply indulge in his affluence by spending time in fashionable and sophisticated Parisian salons. At the age of 35, he gradually withdrew from the life of high society and became more reclusive. While always sickly and seriously asthmatic, he also was a hypochondriac (spending a small fortune on medications) and a neurasthenic. The once flamboyant socialite redefined himself as a hyper-sensitive invalid—and simultaneously began to take his writing more seriously. Proust devoted the rest of his life to writing about his memories of his experiences up until the time he 'retired' from life. The usual English translation for the title of his seven volume masterpiece, *À La Recherche Du Temps Perdu*, is *Remembrance of Things Past*. Proust completed it sixteen years after his self-imposed withdrawal from any sort of social life, dying in 1922 before the last three volumes saw print, but not before the published volumes had received substantial acclaim.

The entire work is in the form of an interior monologue of sometimes excruciating detail.* Although it often deals with elegant society, just as Balzac does in some of the novels in his *La Comédie Humaine*, the two authors could not be less alike. Proust, unlike Balzac, was concerned not with recording the objective world he'd observed, but rather to explore his subjective response to it. Proust's central themes are the nature of memory and the nature of time—and our perception of time. Admittedly influenced by Henri Bergson's theories, and probably Freud's, Proust carried

---

* I should confess I've never made it through the complete work.

introspection to the limits of literary expression. And like the Impressionists, he tried to capture the essence of pure sensation, although not with the immediacy they considered so important, as they painted with quick strokes the too-quick-to-change images before them. For Proust, the embroidered recollection of vivid sensual *memory* of experience was the way to capture the essence of sensation. His most startling success is in the recollection of olfactory memory which, because smell is our most primitive sense, is always richly coloured with emotion.

.

What makes Proust exemplary to a discussion of a study of the limitations of memory and cognition is his methodology, which is based on introspection. Contrast his writing career with that of Balzac. Balzac, like an ardent naturalist, ventured out daily (actually usually nocturnally) to 'collect specimens' from the Parisian social environment. He then would return home with his fresh observations and carefully describe them, sort them, analyze them, and construct a coherent artistic description of his findings. Proust worked, it seems, entirely from old memories. He did not so much study and report on recent findings, as poke around in the dusty mental filing cabinet where he'd tucked away past experience. It is entirely understandable that the nature of time would be one of his central themes, for episodic memory is all about the flow of time.

.

Both Balzac and Proust could be said to have written psychological fiction. What distinguishes Proust is that he constructed his works entirely from exploration of memory, often very old memories, from introspection, from solitary self-analysis, and yet succeeded in creating a fictional world that had enough immediacy and veracity as to continually engage and impassion readers to this day. Memory is not to be trusted as objectively true, but clearly memory can be used, and can be trusted, to create art that is subjectively true and universal.

.

Georg Ferdinand Ludwig Philipp Cantor is a parallel case, for he 'created' a mathematical world inside his head that is far removed from immediate concrete experience, but nevertheless is universally 'true' in the realm of abstraction. The eldest of six children, he was born in 1845 in Copenhagen and raised in a Lutheran German mission in St. Petersburg, Russia. His parents were musical, and he manifested that recurring parallel giftedness in music and mathematics often remarked: he became a highly competent violinist at an early age. His father was rather fanatically religious, and Cantor, despite his intelligence, seemed to never repudiate the religion with which he was inculcated. (He had, like Newton, other strange beliefs that seem contrary to our expectations of those with exceptional intellectual abilities: e.g., he spent a lot of time promulgating the theory that Francis Bacon wrote Shakespeare's plays, even publishing two pamphlets on this wacky idea.)

.

Cantor's family moved to Germany when he was eleven. In 1860, Cantor graduated with distinction from the Realschule in Darmstadt. He then attended the Federal Polytechnic Institute in Zurich, followed by studies at the University of Berlin, where in 1867 he received his Ph.D. for a thesis on number theory.

.

Cantor is most noted for originating 'set theory'. This term is most often associated in the public mind with the infamous misapplication of it in teaching of elementary mathematics. Its real importance is as a major cornerstone of the important domain of 'number theory', and Cantor has been welcomed into the pantheon of great mathematicians. What is interesting in the context of a discussion of creativity is the *nature* of his work, which is way out beyond the borders of conventional thought or anyone's experience of reality.

.

Can you perform arithmetic operations on infinities? If you add two infinities, say the infinity of even numbers to that of odd numbers, is the result a bigger infinity? Is the infinity of real numbers greater than the infinity of just the odd numbers in that set? How does one conceive of one infinity being larger than another infinity?! If you claim questions like this don't make your head spin, you're a damn liar. As Feynman once supposedly remarked about Quantum Mechanics "If you think you understand quantum mechanics, you don't understand quantum mechanics." Well, if you think you can really conceptualize infinity, I dare say you don't understand the meaning of infinity.

.

Cantor is credited with being the first person to 'see' that infinite sets could come in different sizes. Cantor's theorem states that if you have an infinite set of numbers, which contains all possible subsets, the 'power set' of all these subsets has to be greater in size than the original set. It follows from this that there must be an infinite hierarchy of sizes of infinite sets.

If you are reaching the limits of your cognitive ability to visualize this, rest assured you are not alone. I am not a mathematician, but my daughter and several friends are, and they have managed to make a small portion of this clear to me, but I certainly wouldn't even try to explicate it. Cantor had wandered into one of the most *inconceivable* (a most deliberately chosen word) areas of abstract reasoning. And he had returned with a map that is still in use today, and not just by theoretical mathematicians. He corresponded with many of the major philosophers of his time, including Bertrand Russell whom he was supposed to meet in person at a conference, but alas never did. Cantor's influence on philosophy is widely acknowledged.

.

What kind of creative thinking, what kind of cognitive ability, is required to deal with the 'inconceivable' concepts that Cantor played with? No one can even conceive of infinity, never mind the mathematical manipulation of infinities. Well, that's wrong: apparently not *no* one. Cantor could, and now many mathematicians almost casually venture out into the strange territory he first opened for exploration.

.

Any attempt to think about such things as infinity using the tools of conventional language is bound to failure. And what about visualization of infinity? No way. Brilliant mathematicians such as Cantor somehow transcend the limitations of language–based or image-based cognition and then first create—and subsequently take up residence in—worlds that do not and cannot have any place in our everyday 'reality'.

.

But just as living entirely in one's memories, secluded in a cork-lined room, living in the recursively mirrored funhouse of the mathematics of infinity must take its toll, especially when people like your former mentor deride your strange visions.* Some might say that it is just romanticizing what was merely an inherited bipolar disorder to suggest that Cantor's wrestling with ideas in the strange realm of infinities contributed to his eventual mental and emotional collapse into depressive exhaustion. Maybe, but the fact remains that Cantor ended his incredibly productive and creative life as a profoundly depressed man who seemed to have felt for years that he didn't have any emotional or intellectual energy left. Years before his death, he wrote to his friend and colleague Gustav Mittag-Leffler, "how

---

* The highly respected Leopold Kronecker, Cantor's former teacher, did everything in his power to stymie Cantor's academic career, presumably because of a profound inability to accept the 'unreal' realm of mathematics that Cantor was exploring. Kronecker is well known among mathematicians for his statement that "God made the integers; all else is the work of man."

happier I would be to be scientifically active, if only I had the necessary mental freshness!"

.

The great mathematician Hilbert described Cantor's work as "...the finest product of mathematical genius and one of the supreme achievements of purely intellectual human activity." He also proclaimed that, "No one shall expel us from the Paradise that Cantor has created." Most of us find this 'Paradise' a very confusing and frightening place, just as most readers of Proust find the land of his memory as disturbing as it is wondrous. Both are great places to visit, but few would choose to live there—excepting, it seems, people like Proust and Cantor.

LONGING FOR THE INFINITE: LOOKING FOR GOD AND HIS CRONIES

"Be it creative good or destructive evil, if you did it, there must be a reason—or at least so you 'reason'. The human mind is programmed to think of all action as a means to an end."
—Hippokrites

"Life has to be given a meaning because of the obvious fact that it has no meaning."
—Henry Miller (*The Wisdom of the Heart*)

The idea that we give meaning to our existence was given emphatic expression by that fundamental existentialist principle that *existence precedes essence*. We come into existence and then determine what we become.

But of course, what we become is shaped by both what we come into existence equipped—or handicapped—with, as well as by experiences completely beyond our control. Nevertheless, we can't help but feel we continually define ourselves by our decisions.

We base those decisions on a belief that life has some meaning, some purpose. The nihilist may say that life has no purpose, but who can really go on living their life believing that? In fact, many who lose that belief try to take their own lives.

Religion sells a pre-packaged set of answers, but fewer people are buying that shabby product.

For most of us our *raison d'être* is the well being of our family, and for some it may even be the whole human race. Artists and scientists also have to have a reason for what they do.

SPIRITUAL NEEDS AND THE TELEOLOGICAL URGE

The Creationists (or in their latest flimsy disguise, the proponents of so-called "Intelligent Design") are symptomatic of a fundamental human need and the associated drive to fulfill that need, something I'll call the *"teleological urge"*. Only basic biological needs (food, drink, sex) trump the profound human need for 'meaning', for understanding why anything happens, for what ultimately causes and explains human existence. Technically *teleology* is a term referring to the branch of philosophy that studies causes, but it has come to be equated with a specific one of Aristotle's four causes: his *"Telos" or "Final Cause"*; i.e., the reason for which something is done, the goal or meaning behind doing anything. To think that human life has no ultimate cause or goal, to think that life is meaningless, a random event without any *telos*, is a profoundly disturbing thought to most people.

.

Most human beings have satisfied this need for meaning by either embracing a dogma that proffers a formulaic explanation and justification for existence or by creating their own explanation and justification for being. The former include those who are the true believers in any religion or philosophical political orthodoxy. The latter are those who could justifiably be called existentialist in outlook, those who believe that we each *create* the meaning of our lives by living authentically, by fulfilling our potential.

.

But before discussing these two classes of the teleologically concerned, it must be admitted that there *seems* to exist a third group of people who can somehow simply ignore this need—or do not apparently, for whatever reason, even feel it. The outspoken and militant atheist (and brilliant scientist) Richard Dawkins has remarked that he, and most of his highly accomplished and happy friends, do not feel their lives are diminished by not believing in a Divine purpose or in human life being the ultimate goal of some cosmological plan. And I'm sure there are many less distinguished people who live the "unexamined life" Socrates viewed with such contempt, people who do just fine without ever overtly worrying about the meaning of their lives. It is extremely presumptuous to doubt anyone's self-description, especially someone of Richard Dawkins stature, but nevertheless I find it hard to believe he finds his life meaningless. And of course when I call his an 'unexamined life' my tongue is firmly in my cheek. His justifiable fame is based on his unflinching examination of life! His famous book, *The Selfish Gene*, is deeply teleological in nature: it finds the 'meaning' of life in the gene's 'drive' for self-preservation and dissemination.

The problem with 'meaning' is the meaning of the word 'meaning'. Dawkins' 'meaning' is not a particularly satisfying Final Cause for most people, who have grown weary of scientists' continual demotion of their once exalted position in the Grand Scheme of Things. First the earth gets knocked out of the center of the universe, and then even our solar system gets relegated to a backwater suburb at the outskirts of the cosmos. And that damn Darwin and his ilk won't even credit us with a unique place among the trivial inhabitants of our provincial abode out here in Hicksville. Dinosaurs spent more time ruling the roost than we have, and they did not soil this roost as much as we have in our short tenure. But, nevertheless, finding a Final Cause in evolution is still finding a Final Cause. Those who attack evolutionary theory often equate it with the idea that life is meaningless and purposeless, but that is patently not true. It is just evolution doesn't present any evidence to justify raising *Homo sapiens* to the platform of The Final Cause of existence, but evolutionary science, like all science (and art), is still a manifestation of our teleological urge.

So, to continue with my presumption, I do not believe Richard Dawkins really considers his life meaningless and without purpose— or would even say such a thing, despite his (understandable) willingness to mock those who claim one needs a belief in a Supreme Being and Grand Plan if one is to live 'meaningfully'. His refreshingly irreverent book *The God Delusion* devotes one whole chapter to speculation on the persistence and ubiquity of religion (read superstition) and concludes that the most plausible hypothesis is a genetic predisposition to seek meaning—something akin to what I'm calling the teleological urge. As he and the philosopher Daniel Dennett point out, children in their innocence consistently assume that *everything* has a cause or purpose; e.g., rocks are sharp so animals have a place to scratch their itchy backs. Dawkins argues persuasively that this natural tendency is a by-product of a useful evolutionary trait, an inherent heuristic for dealing with the world, which has survival value. The details of this well-reasoned argument can be found in his book or the works of Daniel Dennett.

However despite my deep admiration for Dawkins' hard-headed and brave* analysis of the sources of our teleological urge, and its most irrational and insidious manifestations in religion, I have to carry my presumption further still and say Dawkins is 'in denial'. I offer as irrefutable evidence his own accomplishments. Any one who creates has to implicitly believe that creation has *some* purpose, just as any hard determinist still has to implicitly believe in free-will: one cannot act as if one has no will. His critics accuse him of proselytizing atheism, just as an evangelical Bible-thumper would proselytize his particular variant of Christianity. So when they accuse him of having a mission, they have a point. Yes, the comparison is flawed, for preaching about the virtues of reason is a very different thing from preaching irrationality as a virtue immune to criticism because it is 'faith'. But they are right in saying that he clearly does have a mission. His books have a purpose (and meaning), and he wouldn't have written them if he didn't think it a meaningful, purposeful, way to spend his time and energy.

.

And now to turn to the hoi polloi who supposedly blunder through life with nary even a stray thought about the meaning of existence, those who allegedly never 'examine' their lives. Frankly, I think such folk are few and far between—and largely the invention of self-important intellectuals. Except in those cases where a person's cognitive functioning is really at a level way down the phylogenetic scale, all human beings question the meaning and purpose of their existence. When they fail to find any answer to cling to, they fall into depression. A sure warning sign for a potential suicide is the cry "my life is meaningless'.

.

This brings us back to the two most common paths down which the teleological urge takes us: pre-defined constructions such as religion and existential self-definition. It is tempting for an agnostic or atheist to say that only the latter is associated with creativity, but that is clearly not the case. The former is the road most frequently taken, albeit much more often by artists than scientists.

.

I frequently hear people say that "we all have spiritual needs." I've even had people say to me, believing they are complimenting me, that while they know I "dislike" religion, I'm a very "spiritual guy." My

---

* In a time of increasing religious fanaticism, his willingness to openly mock, satirize or challenge the inappropriate respect and reverence accorded what are clearly crazy beliefs just because they are 'religious' is actually a very dangerous business—as Salmon Rushdie can attest.

response to this is usually a less than gracious denial, for I don't take being called "spiritual" as a compliment—even though I should realize it is intended as praise. Perhaps we all do have 'spiritual needs', if by that is meant that we are all driven by the teleological urge to justify our existence. I'll admit to that. But I think the willingness to find that justification in some dogmatic religious or political system is an unfortunate response to our all-too-human teleological urge.

So, with Dawkins, I have to ask why is this urge so common, even among the creative? It is particularly problematic when it comes to the creative. The creative are creating meaning, so why should they feel any need to search out another source of meaning? This is the subject of the next section, which also deals with creative individuals' attempts to justify and explain their own creativity. But for now, I'd like to consider the existential path.

The primary dictum at the basis of existentialism is that existence precedes essence. This translates into something like we are born (come into existence) and then create the essence of our individual being (essentially pull ourselves up by our own bootstraps). There is no pre-existing meaning to our lives or even life in general: each one of us, individually, creates our self and whatever meaning it has. It is not, therefore, surprising that the concept of freedom is central to existentialism. From a scientific point of view, there are numerous problems with this worldview, which is why the British logical positivists, and philosophers such as Bertrand Russell, engaged in a long and bitter philosophical war with French existentialists such as Sartre.*

On the surface, given any kind of commitment to the principles of empiricism and science, existentialism seems as flakey as religion. Although hard determinism has fallen by the wayside in scientific thinking, no room has yet been made for 'free will' in most of science. And the Lockean 'blank slate' that is implied by saying existence precedes essence seems to indicate inexcusable ignorance of current biological knowledge. But I think these are superficial and overly literal interpretations of the philosophical thinking of those classed as existentialists. The crucial kernel of existentialism is the

---

* It is interesting to note that both Russell and Sartre were awarded the Nobel Prize for Literature. Russell received it in 1950. Sartre was awarded it in 1964, but declined to accept it! His reasons were more polemical and political than philosophical, and many critics have remarked on his longstanding position as apologist for communist atrocities, which seems in contradiction to his existentialism.

belief that meaning, The Final Cause, of every individual's existence has to be found or created within the individual. It is this idea that is relevant here.

.

Artists and scientists do, by necessity, define for themselves the reason, meaning and purpose of what they do. We all do this, at least in some compartment of our lives. Breeding and raising children is a good example. What an absurd thing to do! What personal benefit can possibly be gained by taking on unnecessary responsibility for the care and feeding and nurturing of another human being for, at very least, several decades of one's life? It has been said that the young have children because they are too young to know better. Maybe. The fact is that most of us do it, and most of the time intentionally. And how do we justify it? The good Christian can cite the Biblical injunction to "go forth and multiply". The evolutionary scientist can say it is hardwired 'by' our selfish genes. But let's get real: most parents will admit that having and raising children is one of the things they most feel gives *meaning* to their existence.* Freud and cynics (and perhaps Dawkins) may call this a rationalisation (a made-up *ex post facto* excuse, more than a reason) for a drive we do not understand or consciously acknowledge. That does not matter. Ultimately it is existential. Our children do not give meaning to our lives. We, as parents, are the active agents: we imbue our lives with meaning by freely choosing to have children.

.

So, too, do the creators imbue their lives with meaning by freely choosing to create, not other lives, but instead works of art and science. It is a mysterious process, and sometimes, often, they seem to turn to other than existential self-definition to justify and explain it. This is the subject of the next section.

---

* And if their offspring die or become serial killers, they are among the ones who cry out that "life has become meaningless".

GOD AS HELPER AND THE AMUSING MUSE MYTH

I'm sure I'm not alone in finding it more disturbing than I should to learn that an artist whose work I greatly admire is a member of some absurd religious or political cult or sect. How can a talented singer-songwriter like Cat Stevens convert to fundamentalist Islamism or how can a jazz genius like Chick Corea embrace something as crazy and inane as Scientology.* How is it possible that the great poet Ezra Pound was proud to act as a mouthpiece for the fascists during World War II? Surely they couldn't be finding meaning and justification for their existence—and their roles as creative artists—in such fundamentally repressive and, frankly, morally corrupt dogmatic systems?

.

It is even stranger that most creative scientists, unlike their colleagues in the arts, seem to be largely immune to infection by religious and political dogmas offering up pre-packaged meaning. One would expect them to be *more* susceptible, not less, *because doing science, unlike doing art, cannot be seen as creating meaning.* At least according to the prevailing understanding, scientists *discover* things, not *invent* them,† and what they have found is not evidence of an ultimate *teleos;* on the contrary it seems virtually every new scientific finding presents more evidence of there being no ultimate *teleos*—or at least not one where the individual human life has any importance, no matter how creative or productive that life. What personal satisfaction can there be in finally figuring out even the most fundamental laws of nature, knowing that the whole universe, including this hard-won knowledge, is going to eventually suffer the ultimate heat death where nothing, not even a tiny elementary particle, will ever again move. What possible meaning is there to be found in eternal, unchanging entropy?

.

Yet in the face of such an apparently bleak, nihilistic understanding of existence, scientists seem content to cheerfully just go on with their projects, occasionally taking time out to ridicule the less stoic who flee into the welcoming arms of some implausible dogmatic system promising comfort, promising to give meaning to it all. Were they

---

* One explanation that has been proposed is that when it comes to celebrities, Scientology rewards them well for their endorsements. But that doesn't explain the many who are not celebrities who fall victim to cults.

† At least this is the general consensus. This is not the place to engage the complex philosophical debate about whether scientists *really* discover truth or merely invent plausible metaphors for it.

inoculated with something in their youth to protect them against facile justifications for our existence?

.

I can only speculate about possible explanations for this anomalous difference between the creative scientist and the creative artist. However, before doing so, it is worthwhile to take a more historical perspective on this phenomenon.

.

The great scientists who did the most harm to our 'self-esteem' and self-importance were not atheists. Despite his infamous problems with the Church, Galileo was no atheist, and repeatedly argued that his heliocentric view of the solar systems was not in contradiction to The Scriptures. Newton, alleged unweaver of the rainbow and hero of hard determinists, was extremely devout. He published many more words of Biblical exegesis and scholarship than on his momentous scientific and mathematical discoveries. Newton claimed to read the Bible every day and believed in it as the "Word of God". He even declared his greatest accomplishment was Christian celibacy! Darwin was a regular church-goer most of his life and was known to quote the Bible regarding issues of morality, albeit in his later years, especially after the death of his daughter Annie, he finally repudiated organized religion. Einstein, soon to be discussed as a case study, was famous for his frequent references to God, although his 'spirituality' was certainly not orthodox, and he was basically using the word 'God' as a metaphor for Nature.

.

And of course historically, when it comes to artists, religiosity has always been the norm. Examples need not be cited, for it is exceptions before the Nineteenth or Twentieth Century that are hard to find. From the Muses of Ancient Greece through all the religious art of the Middle Ages and the Renaissance to the mysticism associated with Romanticism and Transcendentalism, some kind of belief in a supernatural justification for—and explanation of— existence and one's creative efforts has been the norm. So perhaps that it still is common shouldn't be entirely surprising, but what *is* surprising is that it still persists despite the fact that all these pre-packaged teleologies of traditional religions—and political substitutes such as Communism, Nationalism, Nazism, Fascism—seem to have been so thoroughly discredited.

.

I offer here a highly tentative hypothesis regarding this issue of the creative individual's degree of need for finding an external *teleos* to explain, justify and give meaning to his efforts and existence. It is based on an assumption that there are levels of *teleos*, just as there are levels of any causal explanation. The *Final Cause* is not something we deem as important as more immediate 'Final' causes. Many a potential suicide gives as the reason for self-destruction the belief that his life in meaningless. But probe deeper: "Why is your life meaningless?" "My only child was my reason for existing, and she has been killed by a drunk driver." Or: "My wife was my life, and she has left me for another man." Or: "I devoted my whole life to my art, and no one thinks it of any worth." It is hard to imagine someone replying: "The whole universe it going to suffer the final heat death, so life is meaningless. Pass me the revolver."

.

The "Final Causes" that matter to us are not final; they are the immediate ones, not The Ultimate One. If a writer is having trouble finding an audience, he can fall back on a belief in the judgment of 'posterity'. That all writers and their works are doomed to eventual extinction when the sun eventually goes nova and the whole universe expands out into an eternal stasis doesn't seem to concern him overly. The belief that gives him hope is that his work will be read for years after his death—which is a mere clock tick of cosmic time. Of course when our more proximate meaning for existence is taken away, the temptation is to reach out to more distal meanings, such as those proffered by religion. The most susceptible to recruitment by cults are those who have lost more immediate justification for their existence: the grieving, the lovelorn, the disillusioned, and the socially disenfranchised. "You find your life meaningless because you have no immediate purpose, no pressing reason to get up in the morning? Well, have I got *just the thing* for you!"

.

The teleological urge or need isn't only satisfied by some contrived Ultimate Answer to the Meaning of Life. It is far more easily satisfied by immediate purpose. It is not to trivialize this to point out that lonely, elderly people are often lifted from their depression by the simple expedient of being given the responsibility of caring for a pet. I offer as a hypothesis that the reason artists seem so susceptible to various, and often dubious, pre-packaged 'purposes' and 'meanings' to life is that more immediate justifications for their existence are wanting. .

.

Artists are vulnerable. Art seems to lack any obvious purpose, so what possible purpose can the creators of it have? Lip service is paid to art's value, but except for the very small percentage of artists who make it big and make big bucks, it is obvious that the majority of people do not consider what artists do particularly meaningful. And even those who are financially successful must realize that they are not being rewarded for something as tangible and apparently deeply meaningful as, for example, a medical discovery resulting in the cure for some disease. Art and Entertainment are, as in the name for the popular TV station *A & E*, equated with each other in most people's minds. But if, like Bach (who dedicated all his works—even light little pieces composed for his children's amusement—to the Glory of God), one can feel that one's art serves a higher purpose—and is not mere entertainment or just the source of a good income—then the teleological need is satisfied.

.

Scientists have it far easier. Ever since the word "scientist" was coined a century and half ago it has had a positive connotation—except of course among those who feel it threatens their own teleology—and theology. It seems intuitive that understanding the world is a completely self-justifying purpose. And if scientific understanding results in better control over the world through technology, how could anyone even suggest it is meaningless or serves no purpose? While obviously scientific advances are not an unqualified blessing (as atomic weapons most clearly demonstrate), it is the rare scientist who doesn't feel confident that uncovering nature's secrets isn't a highly purposeful existence—despite the risks application of such knowledge entails. This immediate gratification of their teleological need seems to leave most of them with no great desire for finding a deeper purpose for their existence.

.

Another thing that should be considered is the differences in the actual creative experience of artists and scientists. The Myth of Muse is based on the feeling that the inspiration, the divine afflatus—that moves the hand holding pen or brush—is somehow otherworldly. Artists often describe feeling not so much like a creator as an instrument being used to create. It follows from the feeling of being an instrument that one is simply serving some greater purpose. The scientist, although frequently reporting similar unbidden flashes of insight, is less likely to attribute these insights to some force outside of him, for he has deliberately worked long and hard toward that particular creative insight. He may credit his subconscious for the insight, but that is still to his credit, for the subconscious is part of his own brain and being.

.

Numerous writers, including such unlikely ones as Henry Miller, have described themselves as "antennae" simply picking up signals from somewhere beyond (from the zeitgeist perhaps, or maybe even God) and merely transcribing them, like a secretary typing up a dictation. This is of course a modern metaphor for the ancient Myth of the Muse. What is surprising and paradoxical is that so many artists are willing to relinquish credit for their own creations, given the conventional stereotype of artist as supreme egotist. Presumably the compensation for not claiming personal credit is to be found in being considered among the 'chosen ones' to transmit 'received' wisdom. Prophets may be revered even more than inventors. And if they are to be believed, there is no doubt they serve a purpose. But to believe one is a prophet is to implicitly posit some external supreme power. Ah, there's the rub. Therein lies the temptation to embrace superstition and religion.

CASE STUDIES: ALBERT EINSTEIN, DANTE ALIGHIERI

Albert Einstein could easily serve as a useful subject for many of the case studies in this book. And it is inevitable that any work purporting to examine creativity in art and science is going to make frequent reference to the scientist who has come to be considered the personification of genius.* It is sensible to be sceptical about such facile hyperbolic labels when they are based on celebrity and public persona, and it is certainly true that the average person's likelihood to name Einstein, when asked for the name of scientific genius, is not based on any understanding at all of what was his real contribution to science. In fact, it was this very inability to understand his theories that increased his stature in the public mind. For decades, it was common for people to say that only a dozen people really understood his theories of relativity, as if the incomprehensibility of something was a validation of the depth of the insight.† It is true that it is a quite small percentage of people (which certainly does *not* include me) who can follow the mathematical and logical reasoning of The General Theory of Relativity, but the basic concepts of The Special Theory of Relativity (which leads to all those famous mind-bending apparent paradoxes, such as the Space Traveling Twin) are understandable to any moderately intelligent person—and *are* understood by any undergraduate physics student—at least any one who is passing his courses. And even the basic ball on a rubber sheet metaphor for his General Theory (that of gravitation) isn't exactly hard to grasp.

'Peer review' is a more valid criterion for judging brilliance than either popularity or inaccessibility, and here the evidence supports the popular conception: Einstein's peers, the other luminaries of physics (those who actually *can* follow with full comprehension his mathematical and logical reasoning), are unanimous in acknowledging his genius. From my reading, I gather the consensus is that probably only Newton is usually credited with greater scientific creativity.

---

* For example, in Michael Hart's book, *The 100: A Ranking of the Most Influential Persons in History,* he is called "the greatest scientist of the twentieth century and one of the supreme intellects of all time"
† Academics seem particularly prone to embracing this dubious assumption, which may explain the reverence accorded complex psychobabble and works by writers such as Derrida.

But the reason I've chosen to consider Einstein as a case study for this section, rather than another, is because of his stature as a radical scientific innovator *and* his public position on the nature of God and spirituality. Consider this famous exchange. Einstein: "God doesn't throw dice!" Niels Bohr*: "Stop telling God what He must do!"

Albert Einstein was born in Ulm, Württemberg on March 14, 1879. His family was Jewish, but certainly not orthodox or even particularly religious. In fact, Einstein was even sent to a Catholic elementary school, presumably because it had a good reputation. He commenced his secondary education at the Luitpold Gymnasium, which was considered 'progressive', but where he, nevertheless, came into conflict with the school authorities over what he perceived as their rigidity and conservatism.† That his parents seemed to take all this in stride suggests that they were anything but religiously (or intellectually) conservative. The remainder of his spotty academic career is well documented, although sometimes exaggerated. Biographers report that the records indicate that, contrary to the legend, he actually was an outstanding student, albeit only in those subjects that interested him—but was, as seems typical of the creative youth, apparently a real pain in the ass for his teachers. It has been suggested that his often abrasive personality, not his inadequate academic credentials, were responsible for him not landing an academic position and having to accept work in a patent office.

It is interesting, and rarely remarked on, that Einstein, unlike his parents, was actually quite devout as a child. While attending Catholic elementary school he insisted on adhering to the practices and restrictions of Orthodox Judaism, which not only annoyed the school officials—but also his own secular Jewish parents. He seems to have abandoned his religious beliefs around the age of twelve. It doesn't appear to have been a sudden about-face, as is so often the case with devout youth who lose their faith. Rather, it seems that his fascination with the beauty and orderliness of mathematics and physics simply redirected his energies, and he just lost interest—and his faith—in traditional religious structures and strictures. Allegedly he was only in his teens when he was already performing the various *Gedankenexperiments* (thought experiments) that were to be the basis for his relativity theories.

---

* Nobel laureate, considered the founder of atomic theory, friend of Einstein, and proponent of quantum theory.
† He later was to praise the institute for its open-mindedness, although it seems that was not his opinion at the time.

Somewhere along the path of his life, Einstein seems to have replaced in his mind the autocratic, arbitrary Yahweh of the Old Testament with a similarly autocratic and arbitrary 'God', which might be considered synonymous with "Nature". He seemed to believe the Laws of Nature were as divine and immutable as the Ten Commandments. Nature passed down these laws to us mere mortals, and it was our responsibility to interpret them and follow them—although, of course, he believed it was inevitable that we will adhere to them. This makes sense of his adamant refusal to accept the idea that God plays dice. Nature is God. God is omniscient and his divine laws of nature irrefutable and immutable. To suggest randomness in the universe is to suggest disorder (which Einstein found repugnant)—and, furthermore, it is a blasphemous heresy. This quasi-religious orthodoxy could well be the explanation for his reluctance to accept quantum theory, even though his own theories were among the building blocks of this theory. It is a similar refusal to accept the role of randomness in natural selection that seems to most annoy Christian fundamentalists. Although their focus is on evolution, not the laws of physics, the major objection raised by proponents of Intelligent Design is virtually the same as Einstein's: "God doesn't throw dice."

This is not to imply that Einstein was 'religious' in any usual understanding of that word or should be put in the same category as Pentecostals. His frequent public reference to 'God' was clearly intended as a metaphor for 'Nature'. But I think it fair to say that his conception of 'Nature' as orderly and rational and even 'just' was how he as a scientist satisfied his teleological urge. Einstein is alleged to have remarked that "If facts don't fit the theory, change the facts." Whether or not he really said that (and I doubt it) he clearly did believe that an elegant, orderly theory was most certain to reflect reality. Again it may be apocryphal, but supposedly his response to the confirmation of his theory of gravity*, years after he published it, was basically a shrug and a remark to the effect that of course empirical evidence had to support such an orderly, elegant theory. If it hadn't, there must have been something wrong with the observations, not the theory.

Scientists who are outspoken in their criticism of religion, such as Richard Dawkins, are often accused of making a religion of science,

---

* The famous observation by Rutherford and others of a solar eclipse that allowed them to precisely measure the degree of bending of starlight passing close to the sun.

just as communists made a religion of their atheistic political philosophy. This criticism, at least regarding science, is bogus and based on a profound ignorance of the nature of science. However, the one grain of truth this criticism does contain is that science, at its very deepest philosophical level, is based on faith in ultimately unsubstantiated (and unsubstantiatable) assumptions—and *faith in these assumptions satisfies a universal teleological need.* Scientists need not turn to religion or mystical beliefs, for science comes packaged with an ultimate teleology: the universe is internally consistent, orderly and inevitable. We all have a part to play in this perfectly orchestrated drama. That our parts may be very minor doesn't detract from the grand overall beauty of the pageant.

.

Dante Alighieri, like Einstein, had this grand vision of the universe. And, like Einstein, he revolutionized his chosen domain. Of course he sought to satisfy his teleological need with theology, not science— for he lived in the Church dominated Late Middle Ages, before science, at least as we understand it, had even been developed as a way of comprehending and apprehending the world. However, like Einstein, his quest for meaning took the form of a grand unifying theory, which he presented in his *Divine Comedy*.

.

Dante was born in 1265 into a prominent Florentine family. He was 'home-schooled', presumably by highly educated private tutors his father could well afford. The traditional story is that he was only nine years old when he met Beatrice Portinari with whom he immediately fell into love—without having even exchanged a single word with her. She was, of course, to become famous as his 'muse', his inspiration to creation. She may have been mortal, unlike the Classic Greek Muses, but Dante having any real interaction with her never compromised his perception of her goddess-like quality. (At most, he exchanged formal greetings with her on the street.)

.

Undoubtedly some understanding of this obsession with a woman never approached can be explained by the then current 'courtly love' concept central to the culture of chivalry. And this phenomenon can in turn be understood within the context of the Christian deification of women, where the Virgin Mary was the archetype of unsullied femininity. The Holy Mother is beautiful, but remains pure and

---

* Philosophers of science readily admit to the problems to scientific veracity presented by Godel's incompleteness theorems, the Kierkegaardean leap of faith involved in inductive reasoning, and the fallacy of affirming the consequent on which so much experimental evidence is based.

virginal, even able to give birth to God's greatest creation, his only begotten son, without any carnality. To this day, the reverence accorded to Christ's mother *inspires* artists and believers. The Madonna and Child is metaphor for The Muse and creation. It is not surprising that until the Enlightenment this was such a popular subject for paintings.

.

Mother Mary being long dead, Dante and the other chivalrous poets projected on to living women the characteristics associated with the Holy Mother. The beauty and—frankly—sexual attraction of a woman induced intense emotional and passionate responses mixed with reverence and awe. The taboo against sullying this demi-goddess would have been very strong. So, to lapse into Freudian jargon, the sexual urge was displaced or sublimated into writing about, rather than ravishing, the object of desire.

.

Lest this all seem an indulging in the same dubious psychologizing I criticize elsewhere, consider the typical modern young man's response to a sexually attractive woman. Even if less than a gifted writer, he'll probably pen some adoring poems or love letters to his goddess. Then, should they or other entreaties lead to courting and mating, or even marriage, almost always the inspiration will dry up. No human being is perfect up close, just as is no work of art is. His goddess will reveal feet of clay. In the best case scenario, his love for her will live on—perhaps even enriched by affection for her foibles—but her status as inspirational deity is inevitably going to be lost. The only way to avoid this is never to bed the one you love, who would eventually wake you one morning with unpleasant snoring or deliver a morning kiss smelling more of garlic than roses.

.

Dante's muse died in 1290 at the pulchritudinous age of 24. At the time of Beatrice's death Dante had been married to Gemma Donati for seven years. Dante had not penned a word about his wife and never did[*], but in 1293 he wrote *La Vita Nuova*, a book filled with poems about Beatrice in which he refers to her as "gracious" and "blessed"—and yet makes scant reference to her physical attributes. However, since he really knew nothing else about her than her appearance, her beauty had to be the unacknowledged source of his inspiration. Even hardheaded sceptics, such as myself, have to admit Freud was on to something with his concept of 'sublimation'.

.

---

[*] Dante's marriage had been arranged when he was mere child, so it is not entirely surprising that his wife was not his inspiration.

After the passing of his Beatrice into memory and personal myth, the great poet turned to philosophical studies and Latin literature to find consolation, reading, for example, Boethius's *De consolatione philosophiae* and Cicero's *De amicitia*. Then in 1340, Dante published *The Inferno*, the first of the three books comprising the *Divina Comedia**.

·

The literary and historical importance of this masterpiece is largely based on its melding and welding of the philosophical and theological with the all-too-human. That it is one of the first major literary works to be written in the vernacular, rather than Latin (the language of scholars and the literati) is part of its importance. That it made concrete the abstractions of theology and philosophy (brutally so in the *Inferno*) is another part. And what makes it of interest in the current context flows from this: It seems Dante conflated his teleological urge with his biological urge in the personification of Beatrice as one of his guides to understanding the structure of ultimate reality.

·

One might say that *The Divine Comedy* is a Middle Ages' religion-based metaphysical version of Einstein's science-based metaphysical conception of an ordered and purposeful universe. Dante made secular a God-based teleology. Einstein made sacred a Nature-based teleology.

---

* This name for the collection of three books, incidentally, was given to them several centuries later.

THE CONSERVATIVE URGE: FADING IMAGES AND SHIFTING PARADIGMS

"It is frequently the tragedy of the great artist, as it is of the great scientist, that he frightens the ordinary man."
—Loren Eiseley (*The Night Country*)

.

"Ignorance more frequently begets confidence than does knowledge: it is those who know little, and not those who know much, who so positively assert that this or that problem will never be solved by science."
—Charles Darwin (*The Descent Of Man*)

As has already been pointed out, one of the attributes that can most confidently be said to be common to creative people is an exceptional ability to see—or draw—relationships between things, often apparently totally unconnected things. The startling, disorienting metaphor is the heart and soul of all great creative insights. It is thus ironic that once such a new connection is seen that new connection blinds almost everyone to any newer connections, even ones that should follow logically from it. Creativity as inevitable precursor of obdurate conservatism seems paradoxical. However, the historical evidence is irrefutable in both science and art. The revolutionaries metamorphose into reactionary establishmentarians almost overnight—just as they always have in the political arena. For some reason, this fact has long been acknowledged in the arts, but scientists seemed to feel they were beyond this jerk-forward and then balk donkey-like progress. Science was a—or at least so scientists claimed—cumulative, smooth progression forward, albeit certainly with occasional exceptional accelerations, but always on track toward a clearly defined destination.

.

Then the historians looked at the empirical evidence, and it didn't seem to support this conception of the enterprise of science as it has actually been conducted. In 1962 Thomas Kuhn published a book entitled *The Structure of Scientific Revolutions* that explicitly contradicted the somewhat smug conception scientists had of how they proceeded. This is only three years after the publication of C.P. Snow's Rede Lecture on the Two Cultures, and the response in the scientific community was similar: indignation. Many scientists felt

Kuhn's book was just adding insult to injury. Understandably scientific self-esteem was very high in the years following World War II. They'd split the atom, demonstrating the awesome power of scientific knowledge. They'd seen their discoveries turned into technology that everyone admitted enriched their lives and into medical advances, such as antibiotics, that pushed life expectancy way up. So what was with this 'scientist bashing'? First Snow claims they are parochial in their knowledge base, ignoramuses when it comes to the arts. Then this Kuhn fellow accuses most of them of being pig-headed intellectual reactionaries, content to blindly accept whatever is the current dogma and hostile to anyone who challenges it. And to top it all off Snow and Kuhn were both respected and credentialed 'real' scientists, holding degrees in chemistry and physics, respectively. They were seen as traitors from within their own ranks.

MORE THAN A NICKEL'S WORTH: BUDDY, CAN YOU SPARE PARADIGMS?

Like C.P. Snow, Kuhn is often misinterpreted, and in fact both eventually felt the need to publicly refute some of these misinterpretations of their theses—by both antagonists with hurt feelings *and* those who wanted to claim them as allies in a war against science. I've already discussed the misinterpretation of Snow. I'd like to now turn to Kuhn's influential work, but with the caveat that I am not free of bias. There is no equivalent of the double blind technique in the interpretation of another's writing, so that good old-fashioned scientific advice 'to go look for yourself' is worth following. Like C.P.'s Snow's *Two Cultures*, *The Structure of Scientific Revolutions* can be read in a few hours, and is very readable.

Kuhn argues that science is not the neat, orderly accumulation of new knowledge through careful, systematic empirical investigation, conducted according to well-established and indisputable rules that guarantee progress. On the contrary, he suggests that a central metaphor and an associated theory inform the daily work of scientists in any field. He calls this a "paradigm" and what most scientists are doing he calls "normal science". As the data rolls in, it has to be made to fit the paradigm, and this is done by "puzzle solving"—a very apt phrase, since like a jig-saw puzzle each new datum is a strangely shaped piece that one has to find where to place it in the grand picture that has clearly defined boundaries, but is incomplete. Eventually a puzzle piece appears that simply does seem to fit anywhere. This is most likely to happen just as the puzzle seems to be nearing completion, for then its failure to fit is only too obvious. The usual response of the scientific community is denial. The oddly shaped piece must be based on bad data, the result of experimental or interpretative error, an anomaly best kicked under the carpet. The scientist insistently waving this piece of the puzzle in other's faces is often ostracized from the community, labelled a crank—or just ignored. But if enough similarly anomalous pieces keep surfacing, or the piece is just too big to ignore, a crisis occurs. And it has always been a crisis of some kind that precipitates a revolution.

So the revolution eventually comes. Someone sweeps the whole nearly-completed puzzle off the table, and replaces it with a new one, a new metaphor, model, theory, or incomplete but coherent puzzle— AKA "paradigm". (For example, Einstein swept classical Newtonian physics off the table, just as the physics community was claiming that only a very few more pieces need to be placed to finish the whole picture.) This is Kuhn's famous "paradigm shift". (This phrase, like Snow's "Two Cultures", soon became part of the lexicon of the philosophy of science.) This is "revolutionary science" as opposed to "normal science".

.

It should be emphasized that the new paradigm is not in any way congruent with the previous one. It isn't that one is entirely starting over from scratch, for all the puzzle pieces, both those that did and those that did not, fit are still around. But in many ways it *is* a totally fresh start, for the boundaries and assumed shape of the completed puzzle have changed dramatically. The profound implication of this is that what once was considered true by virtue of its position in a coherent system is no longer necessarily considered true.

.

This is a much abbreviated précis of Kuhn's thesis, and I have used my own analogies in describing his ideas, but I think I have been fair and accurate in my description of his central concepts. However, I want to be clear that what follows is my own elaboration and evaluation of his ideas and their effect on thinking about scientific creativity.

.

Kuhn was criticized by some (and embraced by others) for suggesting that the scientific theories were not chosen on the basis of reason, and that rival theories were not considered solely on emotional grounds. For example, the older scientists who had devoted their lives to fitting pieces into the puzzle they believed to be the final picture would reject any radical new theory, while younger scientists, less invested in the old vision, would be more willing to go along with a paradigm shift. If this were true, then whatever theory the scientific community held up as true had more to do with the age and personality of the scientists than the actual veracity of the theory! This effectively reduced science to a study in psychology and sociology and implied that scientific truth was 'relativistic'. It was this extreme interpretation of his ideas that Kuhn publicly renounced.

.

Relativism, in the current context, refers to the belief that there are no absolute values or truths or criteria for determining them. To this way of thinking, all of axiology, aesthetics and ethics, is relative. Beauty and ugliness, right and wrong, truth and falsity—all are subjective, probably culturally determined. There is, the extreme relativist maintains, no absolute objective truth regarding what is moral or beautiful or true.

.

Relativism is an unsavoury concept. So are some forms of absolutism, but generally relativists are far more dangerous, anti-intellectual, and *inconsistent* than absolutists. This is because the relativist is actually an extreme absolutist in making the absolute claim that all values are relative. Most absolutists are usually less dogmatic. Most people who maintain, for example, that some behaviours are *really* bad (not just disapproved of in some societies), that some things are *really* ugly (not just disliked by some groups), and that some statements or theories are *really* wrong (not just a matter of opinion) do not claim to be able to make such judgements themselves on everything. Of course some do, but fortunately they are in the minority—or so I'd like to believe. The thoughtful moralist, not blindly committed to some political or religious dogma or authority, will readily admit that many ethical questions are problematic. However, it is one thing to admit that often it is difficult to determine what is the ethical thing to do and quite another to therefore conclude that there is not a more ethical choice.

.

So too in science, it is one thing to admit to the difficulty at arriving at a true scientific theory and quite another to say there are not true scientific theories, only different paradigms accepted at different times. Scientists, even those most invested in their own theories, do not accept the idea that the current dominant paradigm is like the current government in power—in a country prone to periodic revolutions that merely replace one authoritarian regime with a different one. Now, of course, Kuhn did not say this, but many relativists happily claimed he did and were delighted to claim a respected Harvard scientist and scholar as one of their own. One must remember this was the salad days of Derrida and his extreme relativistic philosophy of "deconstructionism". Probably much of the hostile reaction in the scientific community to Kuhn was not so much based on what Kuhn actually said as on what those relativists and deconstructionists gleefully, albeit wrongly, claimed he'd said or implied.

.

What Kuhn really said is generally true, as most historians of science will confirm. I've already outlined the far from orderly nature of creative science as it is actually done. And anyone reading the history of science or even contemporary science news knows that most scientists work within the existing theoretical framework solving puzzles and accumulating knowledge, and it is only now and then some very creative individuals challenge the existing paradigm and cause that framework to collapse—but only because they can offer a larger, sturdier framework. Science *does* progress. Many cynics are quick to point out that the 'idea of progress' is a relatively new one, as if that in itself was evidence of its invalidity. Quantum theory is new; that doesn't; make it invalid. I think most scientists will be among the first to admit that human beings will never progress all the way to some final, complete explanation of the workings of the universe. Understanding has unfathomable depths that may very well be infinite, but that doesn't mean we can't continue to plumb deeper and deeper into these depths. It doesn't mean that, for example, the heliocentric view of the solar system isn't an improvement, a progression in understanding, over the previous belief that earth was at the centre of the solar system, indeed of the whole universe.

Furthermore it is scientists who debunk the idea that evolution itself is a 'progression' that has culminated in *Homo sapiens*. To depict the scientist as one who believes in the perfectibility of Man and expects to achieve omniscience is to set up a straw man—and displays a profound ignorance of the nature of the scientific enterprise.

I do think there are some reasonable criticisms to be made of Kuhn's version of the history of science, but most of them are somewhat unfair for they are all based on his view being an over-simplification, and all generalizations about history inevitably are simplifications. *The Structure of Scientific Revolutions* is not a big book.

One thing it seems to me that Kuhn doesn't seem to adequately take into account are the many unresolved and ongoing 'paradigm skirmishes' that rage in every field of science. Competing paradigms exist side by side for long periods of time and, furthermore, they do sometimes simply merge, as opposed to one completely winning the skirmish and replacing the other in some cataclysmic paradigm shift. A relatively recent example of this regarding theories as to the origin of life might make this point clearer, as well as cast some light on the nature of scientific conservatism.

In 1953 a graduate student named Stanley L. Miller proposed a simple experiment to his advisor Harold C. Urey. Three decades before, two titans of science, Alexander Oparin and J.B.S. Haldane, had (independently) proposed a hypothesis that Miller felt could be put to a simple laboratory test: the hypothesis was that life evolved naturally from complex organic molecules, and that the required organic molecules, these building blocks of life, were created from inorganic compounds present in the earth's atmosphere and sea four billion years ago. The catalyst for this metamorphosis was the extreme electrical activity of the earth's atmosphere in those primordial times.

.

The simulation was fairly simple to perform—a truly lab bench-top experiment. Miller sealed methane, ammonia and hydrogen and water in a sterile glass apparatus containing electrodes. He heated the water to produce evaporation and then sent sparks, as simulation of lightning, through the mini-atmosphere. Then he allowed cooling and condensation, so anything created in the saturated atmosphere by the electricity could dissolve in his simulated primordial sea. After only one week of cycling through this process, Miller and Urey examined their miniature, primitive sea and found that ten to fifteen percent of the carbon in the methane was now in the form of organic compounds, including a significant percentage in amino acids, including thirteen of the same amino acids that are used to make proteins in living cells.

.

Needless to say, their results were big news beyond the scientific community, and the tabloids featured headlines such as "Life Created In A Test-tube!" Of course, they had done nothing of the sort. This was no Frankensteinian experiment, and no living creature had come crawling out of their apparatus. Their artificial primordial soup was entirely lifeless. Nevertheless, their ability to create some of the building blocks of life through replication of natural events that undoubtedly were common on Earth around the time life seems to have first emerged was of profound significance. And Stanley Miller's reputation and stature in the history of science was assured.

.

But here is where the story gets interesting and relevant to the topic of paradigm shifts. Miller hadn't really introduced a new paradigm, for his experiment was based on Oparin's and Haldane's hypothesis and assumptions about the composition of early Earth's atmosphere. But if he hadn't created life in a test-tube, he certainly had given new life to his own scientific career. His fame was assured, and since it was based on this seminal experiment he could not easily go on to question its significance or the theory of life's origins that it purported to support. The remainder of his career was based on elaborating and defending this theory, which he did with progressively more arrogance and acrimony.

Subsequent research cast doubt on the validity of the conclusions Miller was making from this and similar experiments. Perhaps most damaging was mounting evidence that the composition of primeval Earth's atmosphere was *not* what he had assumed it to be and which he had used in his experiment. Although his theory had become accepted as the dominant theory, the current paradigm accepted by the small community of scientists who studied life's origins, it was soon challenged. One of the more convincing alternative explanations followed from the discovery of life in places never suspected, microbial life in rocks deep beneath the Earth's surface, and most significantly in hydrothermal vents on the ocean's floor.

The theorists who proposed hydrothermal vents deep beneath the sea as the cradle of life, rather than the ultraviolet drenched surface of the sea, gradually accumulated substantial evidence for their theory—and were promptly and rudely dismissed by Miller as cranks. They came to be referred to as "Ventists", which like those paradigm-shifters in painting, the "Impressionists", the term was one of disdain. But, like the Impressionists, the Ventists turned the tables and wore the pejorative moniker as a badge of honour.

Miller spent his whole life modifying his theory and attempting to fit new data into it. Of course the Ventists did the same for their theory, although it seems with somewhat less arrogance. And meanwhile other theories emerged, some of which were truly startling, such as the theory that life emerged in rocks deep below the Earth's surface!* All of these theories have supporting data, and what is germane here is the current state of this science of life's origins. Most researchers seem to think the actual explanation is to be found in virtually all of the theories. They do not seem to think the theories are mutually exclusive, except in certain details or exclusivity. So no one paradigm has replaced an older one: instead scientists have discovered ways to reconcile these different theories. The idea that life, or at least the building blocks of life, may originate from very different sources seems to be the current consensus.

So there hasn't been a paradigm shift here, but rather a paradigm merging. I'd suggest this is a more typical event in science than truly revolutionary, sweep-the-table-clean developments such as Kuhn focuses on. Yes, sometimes a major and well-established paradigm

---

* Microbial life has been found a mile below Antarctic ice and in rock bore samples taken from 500 meters below the surface in South Carolina.

has to be discarded. Examples abound, including the mysterious "aether" that allegedly filled space and acted as a medium through which light travels, or "phlogiston", the 'element' in combustible substances that was released when they burst into flame. But often the established paradigm is only wrong in places or is only part of the whole picture, and it gets merged with new paradigms that are similarly incomplete or flawed.

.

I propose that the following scenario is more typical of scientific progress. It, like Kuhn's, admits that science is not a smooth linear accumulation of knowledge, untainted by human frailties. Science does not follow the Baconian model of data collection followed by integration into some grand schema.

.

Kuhn's first period in his scenario is "prescience", a period of observation with some speculation. This is followed by a theory that pulls all these observations together into a paradigm that is accepted by the scientific community and new data are fit into this framework. So far so good. But then troubling data emerge and alternative paradigms are proposed. In the Kuhnian model, these are vigorously, conservatively resisted until a crisis occurs that foments a revolution resulting in the new paradigm replacing the old regime.

.

But often, I would argue, the new paradigm and the old are found not to be totally incompatible, and neither the young heretical radicals with their new vision nor the conservative old guard have their way. Instead integrationists find a way to reconcile the warring parties. These wars probably should more aptly be called tribal skirmishes or family squabbles, for although they can be intellectually violent, these are scientists we're talking about, and so ultimately on the same side—that of discovering truth. Often those not on their side cite these internecine conflicts as evidence of the relativism of science: any disagreement among scientists about the details of the evolutionary process are quickly cited by creationists as evidence that even scientists doubt the theory of evolution.*

.

I think the example from origin-of-life theorists is typical. Miller presents startling empirical evidence supporting a theory in a new branch of science. His theory, being commonsensical, the only one suggested, and one with apparently solid empirical support, is immediately widely accepted. As new data flows in, other theories are

---

* If they were to apply such dubious logic to their belief in God, they'd be in serious trouble. Not much consensus, and much conflict about that theory, I dare say.

proposed. Scientists are fundamentally conservative, and this follows from the scepticism that is central to—and essential to—the scientific method. If it ain't broken, don't fix it. If it works, tell me why I should discard it. And for a scientist who has invested his intellectual life in a particular theory, it is a lot to expect he'll readily admit it was a faulty investment.* Warring camps are set up. Naturally the young, who haven't yet invested in one theory and so have nothing to lose, are more likely to gravitate toward the newer, sexier, more controversial theories. The 'Miller's, armed with an arsenal of experience and strong collegial respect, respond with condescension and mockery. But eventually the passing of its members and erosion from the flow of new data weakens the established camp. Meanwhile the rebellious young Turks mature and concede that there may have been something of substance in the theories of these old guys—and an integration of paradigms occurs.

.

Examples of this abound in the history of science. Psychotherapy is one example. Unlike fifty years ago, most psychotherapists now do not adhere to one school of therapy. More than half describe themselves as "integrative" or "eclectic" in their approach to clients, implicitly conceding some truth in formerly conflicting theories as how to treat their patients. They act as if therapeutic methods (and the theories on which they are based) are just items on a smorgasbord from which they must wisely choose what will best serve their patient.

.

Two things have to be remembered regarding this scenario, which is constantly being acted out in the real world of creative science.

.

One is that the larger paradigm is rarely in question. There are paradigms within paradigms within paradigms. Scientists will often feud bitterly over what to an outsider seem trivial minutiae. But the

---

* I was recently surprised and deeply impressed by a commentator, Jonathan Haas, on a BBC documentary about the extremely ancient Peruvian ruins of Caral. Haas is an expert on emerging civilizations and for decades was a highly respected proponent of the theory that civilizations were always based on fear: warfare forced groups of villages to huddle together for protection, which in turn led to new ways of organizing society, including powerful leaders and division of labor. The archaeological evidence at Caral, a totally unprotected city-state, apparently refuted Haas' theory, and he promptly repudiated it publicly, going so far as to do a BBC documentary. For someone whose whole scholarly reputation and life's work was based on a theory that empirical evidence challenges to so readily admit being wrong is an amazing example of scientific integrity—and more than one can normally or even reasonably expect from mere mortals.

larger paradigm is not in question. All the various proponents of different theories of the origin of life are in agreement that it is a natural event involving the development of organic compounds, followed by those that are specific to living organisms, then onto the first self-replicating microbes, and eventually through the process of natural selection up to life as we know it.

.

The other thing to remember is that while sometimes, as Kuhn emphasizes, a virtually universally accepted paradigm is shoved right off the table by a new one, often there is a war of paradigms that doesn't result in a revolution, but rather in reconciliation and a new paradigm that incorporates elements from both formerly opposing theories. This occurs constantly at both the macro and micro level of theorizing.

.

Science does progress. And it does so at varying rates and along a path littered with discarded frameworks or parts of frameworks that have been replaced by ones that work better. But it *does* progress.

CONSERVATIVE IMPULSES AS A CATALYST TO REVOLUTION

If intellectual conservatism is fundamental to science, as it clearly has to be, how does science manage to continually initiate intellectual revolutions? And is art fundamentally different in this respect? And how does one resolve the apparent contradiction inherent in any form of creativity being in any way conservative? These are important questions that are rarely addressed, perhaps because creativity is so linked in our minds with radicalism, rather than conservatism.

As discussed in the previous section, the scientific method is fundamentally conservative in that existing theories are not easily overthrown—and there is damn good reason for this. No theory becomes established unless there is irrefutable and often voluminous empirical evidence to support it. It simply makes no sense to throw it out on the basis of some extraordinary claim that allegedly contradicts all this accumulated evidence. Extraordinary claims require extraordinary supporting evidence. This is one reason the claims of parapsychologists and other cranks are not taken seriously. Precognition or perpetual motion machines, for example, are contrary to all known laws of physics, and to accept these ideas would require rejecting virtually all established scientific theory. And established scientific theory has served us far too well to be summarily dismissed on the basis of hearsay.

It *is* true that occasionally a single observation can cause a cataclysmic paradigm shift, but it has to be an irrefutable and replicable observation. The most famous example of this rare occurrence is the confirmation of Einstein's General Theory of Relativity by Rutherford's measurement of the degree of deflection of starlight during a solar eclipse observation. But even here, Newton's centuries-old and very reliable Theory of Gravity would not have been replaced by Einstein's had not Einstein also explained the millions of observations supporting the Newtonian Theory. They were damn good approximations that (at a moderate level of accuracy) matched predictions from his General Relativity Theory.

A major feature of conservatism is frugality or parsimony. Parsimony is valued by scientists not just in their theories, but also in their allocation of resources and time. No physicist is going to waste valuable time listening to some bloke who swears he's invented a perpetual motion machine. It is for this reason, that most scientists refuse invitations to debate with creationists. But then, of course, they are accused of being closed-minded and conservative! Well, half right! It isn't closed-minded to choose not to waste one's time. And furthermore, by participating in such a pointless event a scientist inadvertently gives some credibility to such nutbar ideas. As one eminent biologist explained to a creationist he'd declined to debate, such an encounter would look good *only* on the creationist's CV.

.

Nonetheless, it is paradoxical that the scientific method is designed to discourage jumping to conclusions, but nevertheless the major creative scientific insights involve doing just that! Scientists, especially the paradigm shifters, do entertain ideas that are contrary to well-established and reliable theories. I think this is best explained by the perfectionism so typical of creative individuals. If a theory explains something perfectly there is no need to question it or even give it much thought. That ultimately conservative injunction applies: "If it ain't broken don't fix it!" But if it is broken, even in some apparently small and trivial way, it niggles at the perfectionist's mind. At the time of writing, the Voyager 1 is deviating ever so slightly from its predicted path as it passes out of the solar system, and astronomers are publicly speculating if there is a force affecting it that is not accounted for in current theory. They are quick to say that it is probably an undetected technical glitch in the spacecraft guidance system, but what is interesting is their willingness to at least consider an explanation that could require revision of the well-established principles that have successfully guided it 9.3 billion miles from home. This isn't close-mindedness. This is obsessive, perfectionist concern with detail. Everything has to fit perfectly, or else anything goes—including firmly established principles. In other words, the conservative wants everything neat and orderly and is sensibly reluctant to mess with a status quo that is so. But if anything is even slightly out of whack, the problem needs to be addressed. And if radical means are required to do so, so be it. This is how extreme conservatism can act as a catalyst to extreme radicalism. At least in science.

.

How is art different? It is more similar than different. Although artists may on average be less conservative in their daily lives, even their politics, than scientists, they too almost always initially work within the current framework: mastering technique and adopting the contemporary and dominant aesthetic. (Who isn't tired of hearing how talented Picasso was at "traditional" painting and draughtsmanship before he "went crazy".) After long apprenticeship and mounting accomplishment and recognition, why wouldn't even the most talented artist remain 'conservative'? Again, as in science, the answer is an external catalyst—not some inherent contrariness or perversity.

.

What is different in the arts is the nature of the catalyst that initiates radicalism. It is more varied and complex than in science. It can be social or philosophical or technological changes that make the current dominant aesthetic or style seem obsolete. Here is a good example of the secret agent of science (and concomitant technology) infiltrating the artistic community. Beethoven's revolt against the Viennese Classical style was catalyzed by the Enlightenment philosophy*, with its emphasis on individuality, as well as the technical improvements in musical instruments, especially the piano. It is commonplace to remark on how the invention of photography in the mid nineteenth century caused visual artists to search for a new *raison d'etre*—which led to impressionism and then the subsequent bewildering variety of 'schools' of art. And soon afterward the quasi-scientific theories of Freud and Comte led to novelists and poets redefining their roles and goals. Or to cite one of the most dramatic and obvious example: The invention of the printing press forever changed the nature of literacy and literature.

.

Human beings, even the most creative, are naturally conservative. It is the psychological equivalent of the physical principle of inertia. We move along quite contentedly in the straight, unwavering line that follows from some initial shove—until something acts on us to change our direction. It may be that the open-minded, the creative, are just exceptionally responsive to—susceptible to—the pressures of these external forces.

---

* Of course the shift from the Baroque to the Classical was itself also a result of the Enlightenment: the Enlightenment's elevation of the middle class and the belief in accessibility and popularity as virtues in music.

THE SYNCRONICITY OF RADICAL CHANGES IN PHYSICS AND IN ART

Everything keeps on rolling along smoothly like ol' man river—and then bang! Rapids!

.

Punctuated equilibrium is the evolutionary theory that speciation appears suddenly in the midst of long periods of very little change*, and that evolution is not a steady gradual process (the more mainstream theory of phyletic gradualism), but instead that a usually relatively steady state is occasionally and suddenly punctuated by periods of radical development.† One certainly can generalize this viewpoint to history, and specifically the history of the arts and sciences. There are moments in history when paradigm shifts abound and periods when everything remains relatively constant and consistent, changing only ever so gradually. What is interesting is that the periods of dramatic change in the arts and in the sciences so often coincide. The reason is that that even in times when artists and scientists seem to be dwelling in separate domains, they are infiltrating each other's territory.

.

It is obvious and a cliché to say that everything is connected, but nevertheless we don't seem to acknowledge this as much as we should. The increased specialization of creative endeavours, virtually unavoidable because of the exponential expansion of knowledge, is one of the causes of the rift between science and art. It is also why the frequently simultaneous revolutions in very different domains initially may seem inexplicable.

.

In the arts and sciences these punctuation marks have appeared closer and closer together as we travel along the line of history. We still have periods of gradual change and relative equilibrium, but they last for less and less time before some major (or minor) paradigm shift occurs. It has been argued by indignant historians that this is an illusion, and it is certainly true that detecting changes from a great temporal distance is difficult. To lump 1000 years of history into a

---

* The so-called Cambrian Explosion is an example of an *apparently* sudden extreme increase in species after a long period of little speciation.
† The biologist and writer Stephen Jay Gould was the most notable proponent of this viewpoint. Gould's books on evolution include *Ever Since Darwin, The Panda's Thumb, The Mismeasure of Man, Wonderful Life and Full House,* and *The Structure of Evolutionary Theory.*

category called the Middle Ages and view it as a virtually homogenous lump is only superficially reasonable because we are temporally too far away to see the details. However, it is undeniable that change on this planet, in every shape and form, is increasing exponentially over time. And a close look at the curve of this change does support both its increasingly steep ascent—and that this ascent tends to be 'punctuated' by dramatically steep leaps upward.

The musicologist and composer Dan Greenberg presents a striking example of this increasing rapidity of change in an audio demonstration he does in one of his lectures.* He presents two pieces of music, one dating from 700 A.D. (*Ave Marie Stella*, a plainchant hymn) and the other from 1125 A.D. (Thomas of Celano's *Dies Irae*), and challenges his audience to say which is more recent. Because they are based on what were considered the fundamental principles of musical composition during the Middle Ages, only the trained musicologist would have better than a 50% chance of picking the more recent composition. He then presents two pieces of music written within a mere twenty-four years of each other: Rimsky-Korsakov's *Russian Eastern Overture* of 1888 and Igor Stravinsky's 1912 *Rite of Spring* "Dance of the Adolescents". No one with two functional ears could fail to identify the drastically different Stravinsky work as the more recent of the two.

More different visual art movements, more drastic revolutions in the accepted aesthetic, than the previous thousand years, marked the late nineteenth and the Twentieth Century. Even a greatly abridged list is mind-boggling for its diversity.

- Pre-Raphaelites
- Impressionism
- Tonalism
- Symbolism
- Post-impressionism
- Pointillism
- Les Nabis
- Fauvism
- Synthetism
- Art Nouveau

---

* An audio lecture on "How To Understand and Appreciate Great Music" available from the Teaching Company. (http://www.teach12.com)

- Jugendstil
- Modernisme
- Expressionism
- Die Brücke
- Der Blaue Reiter
- Cubism
- Orphism
- Purism
- Futurism
- Vorticism
- Suprematism
- Dadaism
- Neoplasticism
- Bauhaus
- Surrealism
- Les Automatistes
- Constructivism
- Art Deco
- Social Realism
- Abstract Expressionism
- Action Painting
- Color-field Painting
- Outsider Art
- Pop Art
- Magic realism
- Minimalism
- Op Art
- Post-painterly Abstraction
- Hard-edge Painting
- Conceptual Art
- Graffiti Art
- Junk Art
- Psychedelic Art
- Process Art
- Photorealism
- Installation Art
- Deconstructivism
- Maximalism

- Super-realism
- New Realism
- Fractal Art
- Digital Art
- Happenings
- Interactive Art
- Performance Art

Contrast this list with the major art 'movements' of Western Europe for the preceding *fifteen* centuries, where the underlying aesthetic was shared by most artists, no matter how individualistic and distinctive in application of this aesthetic they each may have been.

- Medieval Art (c. 200 to c. 1430)
  - Byzantine
  - Romanesque
  - Gothic
- Renaissance (c. 1300 to c. 1602)
  - Italian Renaissance
  - Northern Renaissance
  - Mannerism
- Baroque (c. 1600 to c. 1750)
  - Baroque style
  - Dutch realism
- Pre-Modern (1750-1860)
  - Rococo
  - Neoclassicism
  - Romanticism
  - Realism

So I think the evidence for an exponential curve is irrefutable when it comes to the arts. What about the sciences? Here too, I can't imagine anyone denying that change has been accelerating at break-neck speed. There probably isn't any need to present a dated, chronological list of major insights into the nature of the universe, but largely because is it fun, here is very brief one that demonstrates the ever increasing frequency of what I think most would agree are some of the most significant scientific breakthroughs.*

---

* Like all lists compiled by anyone of the 'best' or 'most important' anything (books, films, discoveries, art movements, etc.) fault and bias can be found. This list and the preceding lists I think roughly conform to current evaluations by those familiar with the fields. Please just attribute it to my ignorance—not malevolence.

- 1500's
- 1543: Copernicus proposes heliocentric theory of the solar system
- 1543: Vesalius establishes the foundations of modern anatomy
- 1589: Galileo Galilei formulates basic gravitation theory
- 1600's
- 1618: Kepler discovers the three laws of planetary motion
- 1620: Sir Francis Bacon outlines principles of the scientific method in *Novum Organum*
- 1628: Harvey proposes theory of blood circulation
- 1637: Descartes formulates the principles of analytic geometry
- 1665: Robert Hooke discovers cells
- 1684: Newton proposes a universal law of gravitation
- 1687: Newton and Leibniz independently discover the principles of calculus
- 1700's
- 1714: Fahrenheit invents the mercury thermometer
- 1718: Edmond Halley discovers that stars move with respect to one another
- 1735: Carolus Linnaeus develops biological classification system
- 1745: Von Kliest invents the first capacitor, the Leyden jar
- 1747: Benjamin Franklin's kite experiment proves that lightning is a form of electricity
- 1754: Joseph Black discovers carbon dioxide and disproves the phlogiston theory
- 1765: Gaspard Monge invents descriptive geometry
- 1777: Charles Coulomb discovers law of electrical and magnetic attraction
- 1787: Jacques Charles defines Law of Ideal Gas
- 1779: Jan Ingenhousz discovers photosynthesis
- 1789: Lavoisier discovers Law of Conservation of Mass, basis of modern chemistry
- 1796: Georges Cuvier proves extinction
- 1796: Edward Jenner develops smallpox vaccine: beginning of immunology
- 1799: William Smith creates first geologic map

- <u>1800's</u>
- 1800 - Alessandro Volta describes the electric battery
- 1803: John Dalton develops first atomic theory and elemental table
- 1825: Gaus and Loachevsky and Bolyai discover non-Euclidean geometry
- 1827: Georg Ohm formulates Ohm's law of electricity
- 1833: Anselme Payen isolates first enzyme—diastase
- 1839: Schleiden and Schwann propose basis for modern cell theory
- 1842: Julius Robert von Mayer formulates First Law of Thermodynamics
- 1846: William Morton discovers anesthesia
- 1848: Lord Kelvin determines absolute zero
- 1859: Darwin founds evolutionary theory with his book *Origin of the Species*
- 1865: Gregor Mendel works out basic principles of genetic inheritance
- 1865: Pasteur substantiates germ theory by inventing pasteurization
- 1869: Mendeleyev publishes the first version of the periodic table of elements
- 1873: Maxwell discovers electromagnetic theory, placing light in a large context
- 1875: William Crookes invents "Crookes tube" and studies cathode rays
- 1876: Josiah Willard Gibbs founds science of chemical thermodynamics
- 1877: Boltzmann derives statistical definition of entropy
- 1895: Roentgen discovers X-rays
- 1897: J. J. Thomson discovers the electron in cathode rays
- <u>1900's</u>
- 1900: Max Planck lays the foundation for quantum mechanics theory
- 1902: William Bayliss and Ernest Starling discover hormones
- 1905: Einstein develops his Special Theory of Relativity and photoelectric effect
- 1906: Walther Nernst formulates Third Law of Thermodynamics

- 1909: Paul Ehrlich introduces chemotherapy based on his understanding of disease
- 1911: Heike Onnes discovers superconductivity at low temperatures
- 1912: Alfred Wegener discovers the theory of continental drift
- 1913: Niels Bohr publishes his Quantum Theory of Atomic Structure
- 1915: Morgan formulates Chromosome Theory of Heredity
- 1916: Einstein proposes his General Theory of Relativity
- 1919: Rutherford creates first artificial nuclear reaction
- 1921: Otto Loewi discovers nature of nerve conduction
- 1923: Broglie discovers particle-wave duality of light
- 1924: Wolfgang Pauli formulates quantum exclusion principle
- 1926: Schrödinger presents his Theory of Wave Mechanics
- 1927: Heisenberg formulates the uncertainty principle
- 1927: Georges Lemaître proposes: Theory of the Big Bang
- 1928: Paul Dirac formulates quantum mechanics
- 1928: Linus Pauling describes nature of the chemical bond
- 1928: Fleming discovers penicillin, the first antibiotic
- 1929: Edwin Hubble defines Hubble's Law regarding expansion of the universe
- 1932: Ernst A.F. Ruska and Max Knoll build the first electron microscope
- 1937: Alan Turing describes a theoretical precursor to the modern digital computer
- 1943: Oswald Avery proves that DNA is the genetic material in the chromosome
- 1947: Shockley, Bardeen and Brattain invent the first transistor
- 1948: Claude Elwood Shannon invents Information Theory
- 1953: Crick and Watson describe the helical structure of DNA—basis for molecular biology
- 1957: USSR launches first artificial satellite
- 1958: Precursor of the Internet (ARPA) created
- 1960: Perutz, Rossmann, Cullis, Muirhead, Will, North describe the structure of proteins
- 1960: Allan R. Sandage discover quasars

- 1964: Penzias and Wilson find confirming evidence for the Big Bang
- 1965: Richard Feynman develops Quantum Electrodynamics
- 1965: Leonard Hayflick discovers normal cells division limit—The Hayflick limit
- 1967: Jocelyn Bell Burnell and Antony Hewish discover first pulsar
- 1967: Steven Weinberg presents a unified theory of forces, *sans* gravity
- 1969: Neil Armstrong and Edwin "Buzz" Aldrin land on the lunar surface
- 1971: Ted Hoff creates first microprocessor
- 1972: Interstellar space craft (Pioneer 10) launched
- 1975: Edward Roberts invents first personal computer
- 1978: First "test tube" baby is born
- 1978: First computer spreadsheet (Visicalc) created
- 1984: Kary Mullis invents the polymerase chain reaction, a key to molecular biology
- 1989: Internet use increases dramatically and with it universal information access
- 1995: Mayor and Queloz observe the first extrasolar planet
- 1995: Fermi Lab physicists discover top quark, further confirming Quantum Theory.
- 1996: Wilmot and Campbell clone a sheep—Dolly
- 1999: Human Genome Project begun

Well again this list too should boggle the mind. I used to wonder at the scientific and resulting technological advances my mother had been privileged to observe in her lifetime. Cars were still a novelty when she was a child, and she lived to see human beings walk on the moon. And now when I look back at the changes I've seen in my own lifetime, I feel like I've lived through several aeons of radical changes.* No peasant of The Middle Ages experienced any change in his world approaching what my mother and I have witnessed. It dates me, but I really remember when television was new and a novelty. Young people now have no conception of what the world was like a mere few decades ago: a world without computers and the Internet?

---

* It's been a wonderful roller coaster ride, but it makes me feel old. Hell, I remember when there were no home computers, and even when TV's came out! I was privy to seeing the first broadcast television shows, including the classic "Kookla, Fran and Ollie".

Inconceivable! Frankly, *I* find it inconceivable, and I was there when it all began.

.

But to return to the issue of the apparent synchronicity of paradigm shifts—or revolutions—in art and science, one has to try to evaluate the likelihood of such a correlation being causal or just casual, i.e., virtually accidental. Historians and philosophers talk about the *zeitgeist*—ideas 'in the air' at particular times in history that influence people in every walk of life, from politicians to scientists to artists to the average person. Certain events, be they new ideas or political upheavals or discoveries, can exercise an influence that is so pervasive it catalyzes change in virtually every current domain of human endeavour.

.

For example, it is undeniable that the French Revolution and Napoleon's military adventures and misadventures had a profoundly catalytic effect on the arts. Infatuated, like many artists, with Napoleon, Beethoven dedicated his Third Symphony to Napoleon—and then in disgust at his allegedly egalitarian hero's declaring himself emperor, tore up the dedication. And it was unquestionably advances in science that help trigger the Romantic Movement in literature—as a reaction against what was misinterpreted as the dehumanizing effects of science. The horrors of World War I do much to explain art movements such as Dadaism. To reiterate the cliché: everything is connected.

.

There is a truly wonderful book by the surgeon Leonard Shlain entitled *Art & Physics: Parallel Visions In Space, Time & Light*. Being neither professional physicist nor artist, he is better qualified than either specialist to see the parallels in these two domains. (This is not to imply he isn't knowledgeable and enthusiastic about both physics and art, for he is—perhaps more so cumulatively than many of those who actually earn their daily bread in one or the other of these activities.) Although sometimes the connections he draws between revolutions in scientific perspective and artistic perspective may be subject to sceptical questioning, overall the evidence he mounts for the subtle, largely unacknowledged, effect scientific revolutions have had on artistic revolutions make him one of my corroborating witnesses for the testimony of this book: there are secret agents, unnoticed, at work in the two camps of science and art.

.

From around the middle of the nineteenth century until the beginning of World War I, Paris had more cafes and drinking establishments than any other city in the world*, and it was the intellectual and artistic centre of the Western world. It is fascinating to fantasize about many of the great creative minds in math and science and the arts sitting at adjoining tables in some outdoor café, emanating ideas from their different domains that waft through the air, mingling, and then drifting back down onto all the tables. It's a nice metaphor for *Zeitgeist*.

* For those who think their city was more lively, I'll give my reference for this claim: "Cafe Friend': Friendship and Fraternity in Parisian Working-Class Cafes, 1850-1914" by W. Scott Haine in the *Journal of Contemporary History*, Vol. 27, No. 4 (Oct., 1992), pp. 607-626.

CASE STUDIES: BLAISE PASCAL, EZRA POUND

Two striking examples of the paradoxical nature of conservatism within creativity are the Seventeenth Century mathematician, physicist and philosopher, Blaise Pascal, and the Twentieth Century poet, Ezra Pound. Both revolutionized their respective domains, and both were deeply, even irrationally conservative by nature.

Blaise Pascal was born on 19th June 1623 in a provincial France. When his mother died three years later, his father, Étienne, moved his three children to Paris where he took on responsibility for their education. As is often the case with the highly creative individuals, especially in the fields of math and music, Blaise Pascal's gifts manifested themselves early. At the age of eleven he wrote a sophisticated treatise on vibrating bodies. Interestingly, his father did not initially encourage his son's pursuit of mathematical and scientific interests but actually discouraged it for fear it would detract from his more classical studies. However, eventually Étienne relented and his son was allowed to study geometry and observe discussions at gatherings of eminent mathematicians and scientists organized by the monk (and mathematician) Père Mersenne.

At the age of sixteen, Pascal wrote his first mathematical treatise entitled an "Essay on Conics" which now is known as Pascal's Theorem. When Mersenne showed the work to Descartes, Descartes initially insisted it had to be written by the father, not the son, for no child could have created it. At age of eighteen he invented a mechanical calculator capable of addition and subtraction, a very early precursor to the computer, but treated at the time as an amusing toy. It was and still is called Pascal's Calculator or the Pascaline. Throughout his twenties he continued to make major contributions in areas of science, including fluid dynamics, the nature of pressure and vacuums, as well as to various fields of mathematics, including projective geometry and probability theory. Perhaps the most important of his many contributions was to the philosophy of mathematics, including the nature of axioms and definitions and proofs. Many of his insights in this area were not appreciated till a century after his death, because of the delayed publication of his *De l'Esprit géométrique* ("On the Geometrical Spirit").

Famous and respected as the name Blaise Pascal has become among mathematicians and scientists, the average educated person probably knows him best for his posthumously published literary and theological work, the *Pensées* ("Thoughts"). And particularly for his famous Pascal's Wager, where he argues that belief in God is the only safe bet, for no matter how small the odds are of God's existence, the results of disbelief are disastrous—while the results of belief, even it is wrong, involve no loss!

.

It is reasonable to ask how this religiosity fits within his scientific and mathematical accomplishments. How could such a revolutionary thinker be so religiously conservative? The answer is to be found in his biography, and it is interesting to see how he attempts to resolve the cognitive dissonance of these two sides of his personality.

.

Pascal's health was never good, and he was in constant pain from one ailment or another from his adolescence on. He was also emotionally over-sensitive. Math and science may have served as intellectual balm for his torments. And then he found another salve.

.

Allegedly in November of 1654, at the age of thirty-one, Pascal was in a carriage accident where the horses plunged off a bridge as the reins fortunately snapped, the coach was left hanging halfway over the edge of the parapet. Pascal and the other occupants of the carriage were rescued, but shock of this near death experience caused him to faint. The story is that after fifteen days of virtual delirium and shock, Pascal awoke from a dream that he interpreted as a mystical experience. He promptly scribbled a note about the experience that he subsequently kept secreted in the lining of his coat and which was only discovered after his death. Here is an English translation by Elizabeth T. Knuth of this note, which has come to be called "Pascal's Memorial".

.

FIRE.

> GOD of Abraham, GOD of Isaac, GOD of Jacob
> not of the philosophers and of the learned.
> Certitude. Certitude. Feeling. Joy. Peace.
> GOD of Jesus Christ.
> My God and your God.
> Your GOD will be my God.
> Forgetfulness of the world and of everything, except GOD.
> He is only found by the ways taught in the Gospel.
> Grandeur of the human soul.
> Righteous Father, the world has not known you, but I have known you.

Joy, joy, joy, tears of joy.
I have departed from him:
They have forsaken me, the fount of living water.
My God, will you leave me?
Let me not be separated from him forever.
This is eternal life, that they know you, the one true God, and the one
that you sent, Jesus Christ.
Jesus Christ.
Jesus Christ.
I left him; I fled him, renounced, crucified.
Let me never be separated from him.
He is only kept securely by the ways taught in the Gospel:
Renunciation, total and sweet.
Complete submission to Jesus Christ and to my director.
Eternally in joy for a day's exercise on the earth.
May I not forget your words. Amen.

.

It clearly reveals that the Catholicism he had up to this time been viewing with such disinterest and even scepticism had left its mark on his sensitive psyche, and so in his emotionally distraught state, he was converted. The saying that there are no atheists in the foxholes isn't true, but there is no question that fear of death has been a major catalyst to religious conversion. Pascal's near-death experience changed him dramatically and permanently.

.

His sister Jacqueline had several years before converted to the fundamentalist and strict Jansenist sect whose followers believed that the dominant and traditional Jesuit Catholicism of the time was lax and decadent. Their father's doctor had introduced both Blaise and his sister to Jansenism. After Pascal's conversion, it was the tenets of this harsh—sometimes labelled heretical—sect that he embraced.

.

He gave up his scientific pursuits, and most of his mathematical investigations, to devote himself to his newfound, highly conservative brand of Catholicism—and to attacking the traditional Establishment Catholic church for its lax morality. He immediately revealed another talent: a literary one. Two years after his conversion he published (under a pseudonym) *Provincial Letters*, a fictional report from a sophisticated Parisian to his country friend about the moral and theological issues embroiling the intellectual elite of Paris. It was a brilliant and biting satire of Jesuit casuistry, an early form of situational ethics that born-again Pascal found abhorrent. The book was banned, shredded and burned—but not without reaching a wide audience. Pascal continued his theological and philosophical writing until his death eight years later at age thirty-eight, with his master work the *Pensees* not quite complete and only edited for publication after his passing.

That he completely embraced the rigid moral absolutism and asceticism of Jansenism is evidenced not just by his writings. During the last years of his brief life, he was reputed to have refused attempts to ease his suffering, saying "Sickness is the natural state of Christians." Since his life was one of constant suffering, and he had become a very serious Christian, this awful connection and conclusion isn't entirely surprising.

It is interesting that Pascal's admirers included not just other religious conservatives, but also such unlikely fellows as the agnostic Voltaire and the radical Jean-Jacques Rousseau. (Surely Pascal would not have reciprocally admired either of their philosophies.) The explanation is partially that Pascal wrote so very well, and the French have always had the greatest respect for literary style and eloquence. The other part of the explanation may be that they all had a common enemy: the Jesuit-dominated Catholic Church of France at the time.

How is it possible that Pascal, the great innovator in mathematics and science, could metamorphose into such an intensely conservative and doctrinaire Christian that even the obviously conservative Catholic Establishment offended him as being lax and libertine? There are two things that make this less paradoxical.

First, it could be reasonably argued that he wasn't being conservative at all, for he wasn't defending the established Church: he was attacking and mocking it. The Establishment no doubt considered him as radical as Voltaire. Conservative is an ambiguous term, as its use in politics shows. (Were the Bolsheviks, once they came into power, conservative or radical?) In the context of a discussion of creativity, I have used it to mean being reluctant to change from the established aesthetic or paradigm—without perceived justification. And I have argued that this is a common feature of the creative scientist, and even the creative artist, *but* with them it is paired with a willingness to admit justification when they encounter it—and then make a dramatic and radical break. Surely this applies to Pascal. It may seem to many, myself included, that there wasn't real, rational justification for his conversion, but certainly from his perspective there was.

.

Second, there is a conservative consistency in his views and attitudes that is masked by the contrast of religious belief and commitment to scientific investigation. In both domains Pascal was very much a hardheaded sceptic and absolutist. I can't claim the mathematical sophistication to fully understand his analysis of definition and the limitations of deductive reasoning, but I gather this is considered one of his major contributions to the philosophy of mathematics. And while a staunch defender of the scientific method and inductive reasoning, he also pointed out its limitations and unsupportable a priori assumptions. Once he had converted to his brand of Christianity, he tried to apply the same conservative logical rigour to his theological thought. It was his interest in probability theory that led him to his famous wager as a justification for belief. Given the a priori equating of religion with the Christian religion and its beliefs about the afterlife, this often-mocked bit of reasoning makes some sense. (Of course the flaw in it is the assumption that there are only two options, two possible bets: atheism or some specific brand of Christianity. As Richard Dawkins points out, say you place your bet on the Christian god, and then you end up being greeted at the Pearly Gates by Baal or Zeus or Odin? Not good!) And certainly it isn't a traditional 'conservative' Christian argument to suggest belief in Yahweh should be based on probability theory and self-interest!

.

Pascal's genius is indisputable. His contributions to science and philosophy and literature are stellar—and his beliefs not as paradoxical as they may seem.

.

And then to race forward almost three centuries, consider another paradoxical figure: the poet Ezra Weston Loomis Pound. Born in Idaho in 1885, Pound is considered by many critics to be *the* seminal figure in literary modernism—not just because of his own writings, but also because of his personal influence on so many of the other key figures that reshaped literature. Most noted, of course, is his editing of—followed by promotion of—T. S. Elliot's radical long poem, "The Wasteland", but his aesthetic views influenced many other important writers, and his promotional efforts on behalf of writers we now consider the earliest 'Modernists' changed the literary landscape forever: e.g., James Joyce, Wyndham Lewis, William Carlos Williams, H. D., Marianne More, Robert Frost.

.

However, he is also infamous his anti-Semitism and for his support of Fascism during World War II. Objective evaluation of his work has been hampered by his reputation as a political reactionary of the worst kind.

Pound studied at the University of Pennsylvania—where he met Hilda Doolittle (H. D.) and William Carlos Williams. He went on to earn an M.A. in Romance Philology from Hamilton College in 1905. He taught briefly at Wabash College in Indiana, but got into trouble for bringing a woman to his room—an actress, no less! In 1908 he moved to London where he met the literary greats W. B. Yeats and Ford Madox Ford. Like many artists of the time, the horrors and absurdity of World War I profoundly darkened his view of human nature and coloured his art and his thinking with a pessimistic nihilism. In 1920, after extensive travels in Europe, he moved to Paris, where once again he became friends with artists who were seminal figures in Modernism, including Marcel Duchamp, Tristan Tzara, and Fernand Leger. In 1924, Pound left Paris and moved to Italy, where he finally settled down. It is here he met James Laughlin who he inspired to start what I consider one of the most important literary publishing house of the Twentieth Century: New Directions.*

Along this peripatetic way, Pound managed to publish several books virtually every year. They include his brilliant and controversial 'translations' from the Japanese inspired by Ernest Fenollosa and many volumes of his own verse.† Throughout this time he was also working on what was to be his masterwork: *The Cantos*.

However, his literary accomplishments are well known but not what is of relevance here. Rather, it is his infamous support of fascism after the outbreak of World War II and the apparent paradox of such a revolutionary figure in the founding of Modernism being so politically reactionary.

When World War II began, Pound remained in Italy. He, strange as it seems, found Mussolini and his Fascist philosophy appealing. When

---

* When I was growing up in Chicago, there was a large Kroch's and Brentano's Bookstore downtown that for some reason sorted their books by publisher. Whatever literary education I have, I attribute to discovering the New Directions (and Grove Press) sections, where any random selection was bound to be an exciting read—and one very different from the literary canon presented in the formal educational system.

† He also was composing music. Like Nietzsche, another master of the music of language, Pound actually composed music that is still of interest.

the United States entered the war in 1941, Pound interspersed his literary work with writing Axis propaganda. Italian radio gave him a platform to talk about 'culture', a platform from where he conflated 'culture' in the aesthetic sense with 'culture' in the sociological/political sense. He was wise in the former, and deluded in the latter. Having become obsessed with money and economics, and clearly tending toward paranoia, he seems to have bought into the conspiracy theory that Jews, especially those involved in banking and money lending, were responsible for all the economic woes plaguing Europe and the United States. Some of the nasty, crazy things Pound said on radio and wrote at this time are even more damaging to his reputation as a brilliant thinker than Pascal's post-conversion writings are to his reputation as an enlightened individual. It is profoundly important to acknowledge that great, creative minds—even those whose creativity spans various domains—can be almost unbelievably *stupid* about things that, for whatever reason, tap into some short-circuit in their emotional system.

.

At the end of the war in Europe, Pound was captured by Italian partisans, and eventually handed over to U.S. forces and sent back to the United States where he was tried for treason. He escaped conviction of this capital offence by being found legally insane and sent to St. Elizabeth's Hospital in Washington, D.C., where he was to remain incarcerated from 1946 to 1958. Whether his alleged insanity was a plea bargain to avoid the possibility of a death sentence, or whether it was in any way justified by psychiatric evidence of serious mental illness, is still debated. His continued literary output and the testimony of friends who visited him at St. Elizabeth's do not support the contention that he had gone stark, raving mad. Having some crazy ideas does not make one clinically crazy.

.

So how to reconcile Pound's reactionary political views with his revolutionary artistic views? As with Pascal, there is a subtle intellectual consistency beneath this apparent contradiction. Pound was in many ways a classicist whose 'revolutionary' ideas regarding literature were largely drawn from traditional sources in other cultures. Imagism, the school of poetry he virtually founded, is based on Japanese and Chinese models where richly connotative images (called "pillow words" in Japanese haiku and tanka) carry the weight of the poem. The poetic meter used by his protégés such as Marianne Moore is syllabic, again the meter of classical Japanese verse. Breaking from the English tradition of accentual-syllabic verse was not so much revolutionary as a harkening back to classical models, albeit from a different culture. Compare Pound's famous poem "In A Station of The Metro"...

> The apparition of these faces in the crowd;
> Petals on a wet black bough.

or William Carlos Williams equally famous and terse imagist poem "The Red Wheelbarrow"...

> So much depends
> upon
>
> a red wheel
> barrow
>
> glazed with rain
> water
>
> beside the white
> chickens.

or with this poem ("In The Quiet Night") written in the Eight Century by the Chinese poet Li Po*

> The floor before my bed is bright:
> Moonlight - like hoarfrost - in my room.
> I lift my head and watch the moon.
> I drop my head and think of home.

or with virtually any major Japanese or Chinese short lyric poem from a thousand years ago.

Pound was merely grafting a great poetic tradition onto the tradition of English literature. It is almost superfluous to talk about 'influence'; Pound included translations into English of Li Po's poems in his acclaimed and controversial book of translations, *Cathay*.

Pound's economic theories, also based on the earlier traditions, might actually be considered revolutionary in the same way his poetic theories were, for they were an attack on the current status quo. They may have been viewed differently had not anti-semitism and their association with fascism not tainted them. He opposed income tax

---

* Translation by Vikram Seth.

and governmental spending where the charging of interest inevitably produced an ever expanding and out-of-control national debt. Although some of his fiscal ideas may have been naïve, many contemporary, so-called Red Tories (i.e., fiscal conservatives who do not have conservative views regarding social reform or tolerance of diversity or artistic innovation, but who are deeply concerned about the pernicious influence of both Big Business and Big Government) would agree with the general tenure of his argument—sans, of course, blaming Jews for it all.

.

With Pascal, it could be said that he carried over much of his early radical rationalism and empiricism, which included the scepticism that is often conflated with conservatism, and tried to apply it to the irrational domain of religion. But even there he was opposed to the status quo. With Pound, his innovations in the arts seem to have been inspired by his respect and admiration for unappreciated global traditions, not by a radical rejection of tradition. And his ultra-conservative politics, misguided as they became during the war years, were based on the same scepticism of the values of contemporary society that one usually associates with radicalism.

.

Both Pascal and Pound even in their support of 'conservative'—and what seems to us backward and anti-liberal causes—were outspoken in their opposition to the status quo, their scepticism of its value, and their belief in a radical change in current values. This is always part of the creative personality: disagreement and rejection of the current view of things. What is offered as an alternative is not always right; it may in fact be silly or even result in evil. Change isn't inherently good, anymore than stasis is, but creativity *is* based on changing things when it seems things need changing.

.

Pascal and Pound both demonstrate how complex the relationship is between the sceptically conservative aspect of creativity and the radical, revolutionary side—in both scientist and artist. Which one drives the other varies even within the same individual. That the driver steers down the right road isn't assured. One thing that can be said is that no matter how 'conservative' the creative may seem by one or another definition of that slippery word, they do not want to just park by the side of the road.

DANGEROUS SEDUCTIONS: THE PSYCHIC CIRCUS

"Now my own suspicion is that the universe is not only queerer than we suppose, but queerer than we can suppose."
—J. B. S. Haldane ("On Being the Right Size" in *Possible Worlds*)

"A man is his own easiest dupe, for what he wishes to be true he generally believes to be true."
—Demosthenes (Attributed)

Successful and dangerous seducers disguise their real intentions, and masquerade as friends. A well-established attribute of psychopaths is their charm and ability to convince people of their sincerity. They use the tools of flattery and promises to get what they want. And it is our entirely natural tendency to prefer praise and hope over honesty. In fact, honesty and commitment to truth is so resented that it is often punished.

Truth is illuminating, but sometimes it can be a harsh light. But it allows us to see more clearly despite our imperfect vision. And that is to be preferred.

THE FALSE MYSTERIES: THE BANALITY OF THE SUPERNATURAL

As if the bizarre counter-intuitive paradoxes of Relativity Theory and Quantum Theory weren't enough to stretch our imaginations and inspire awe, a recent theory that tries to resolve the apparent conflicts between these two fundamental physical theories involves postulating eleven physical dimensions and two *time* dimensions!* With such an Alice In Wonderland universe out there for exploration, one really has to wonder why people are so fascinated with such dull products of the human imagination as ghosts or other so-called 'paranormal' or 'supernatural' phenomena. Richard Dawkins has given a wonderful lecture entitled "Queerer Than We Suppose: The Strangeness Of Science" in which he points out the awe inspiring mysteries of the real natural world and wonders at why anyone should be drawn to the far less awesome, far less interesting, unreal imaginings proffered by proponents of the so-called 'supernatural'. Nature *is super*—nature *is* super-natural. What are called 'supernatural' phenomena pale by comparison.

There are a number of excellent books† debunking the 'paranormal' while simultaneously revelling in the wonderfully strange nature of the so-called 'normal' world, so there certainly is no need for me to do so. It may be a rash assumption in a time when polls indicate 60% of the U.S. populace believes in the literal truth of *Genesis*, but I will assume that at least my readership does not need convincing of the absurdity of clairvoyance, telepathy, water-divining, communication with the dead, telekinesis, astrology and all the other superstitious nonsense that seems to have such wide appeal. However, it *is* relevant to my theme to ask why those creative individuals working in the arts, and who are obviously very intelligent, seem to be almost as gullible as the ill-educated who visit 'psychics' and believe in ghosts and chatting with one's dead grandma at séances. They should know better. Or so one would think.

But clearly many don't. Examples are rife, but just to mention one unlikely individual—Sir Arthur Conan Doyle was a devout, even

---

* Itzhak Bars at USC college.
† Books by Carl Sagan, Martin Gardner, and James Randi come to mind: *The Demon-Haunted World : Science As a Candle in the Dark* by Carl Sagan; *Did Adam and Eve Have Navels?: Debunking Pseudoscience* by Martin Gardner; *An Encyclopedia of Claims, Frauds, and Hoaxes of the Occult and Supernatural* by James Randi.

militant, spiritualist.* Now this is the creator of Sherlock Holmes, that fictional paragon of reason! And Conan Doyle was not a wild-eyed romantic, ignorant and fearful of science. He was trained as a doctor and quite interested—and knowledgeable—about the science of his time. Even his fictional hero, who is so oft cited for his *deductive* powers, actually more often used scientific, empirical *induction* to solve crimes.† So how explain a belief in spiritualism and communication with the dead in a writer who seemed to be bridging the two cultures?

.

I think the explanation at least partially lies in the old maxim that "a little knowledge is a dangerous thing." Then combine this with the arrogance bred of demonstrated intelligence and accomplishment in another creative field. Finally stir into the mix those factors that are virtually universal in their tendency to mislead people into the most implausible of beliefs, and you have something very potent—and intellectually toxic.

.

What are these factors? This is my short list, but I am sure there are others.

- Only coincidences we see as meaningful are remarked
- Misperceived order
- Incomplete understanding of probability
- Projection
- False memories
- Desire to believe

.

What follows is an attempt to consider these factors in the specific context of the creative individual, for there very well might be something about the gift of creativity that makes the creative individual particularly susceptible to their influence.

.

Only Coincidences We See As Meaningful Are Remarked

.

The number of 'coincidences' in a single day is virtually infinite. For example, right now as I am typing, my dog happens to be sleeping in the bathroom, wrapped around the toilet bowl. What a coincidence! Here I'm trying to write prose that isn't soporific and not likely to be flushed away in the next editing, and here is my dog by his insensitive actions implying that I'm failing terribly. Now you might say I'm

---

* The famous debunker of spiritualism, Houdini, Sir Arthur viewed as an arch enemy. Some conspiracy theories even try to implicate Doyle in Houdini's sudden death!
† Just one example is Sherlock's taxonomy of tobacco ashes.

being paranoid, seeing a meaningful relationship between two simultaneous events when there is none—and I certainly hope you are right. You might go on to say this is just a random coincidence, but wait: here there actually may be some connection. It is quite hot today, and the lure of leaving my desk to sit outside on the deck is reduced, so I am writing instead. The heat is also the reason my Black Lab has decided to wrap himself around the cool toilet bowl—something he does whenever it gets too hot in our un-air-conditioned house. These are more plausible explanations for this particular coincidence than assuming he is responding telepathically to my writing.

However it is also true that I just heard on the radio, which I have playing softly in the background, that there was another homicide in Toronto, which I am reasonably sure has no connection with my murdering the English language. So this I think I can reasonably attribute to pure chance. Coincidences abound. Things are constantly happening simultaneously, and we only remark on those where we see or imagine a relationship.

If, as I have argued, what distinguishes the creative intelligence is seeing or imagining relationships between apparently unconnected things, then it is hardly surprising that artists are especially susceptible to finding more things remarkable. This is, after all, what metaphor is all about. Scientists, on the other hand, are trained (or naturally inclined) to go beyond merely connections—to go on and examine all the coincidences they find remarkable for some rational explanation. (The remarkable coincidence that a person in a free-falling elevator and a person in stable orbit over the earth experience the same weightlessness was one of Einstein's inspirations for his gravitational theory.) In very modest form, the above tale (tail?) of my dog and my writing exemplifies this difference in response to natural hypersensitivity to seeing relationships. Imagining my dog commenting on my writing is typical of artistic creativity. Going further and trying to find some plausible rational connection between the two events is typical of scientific creativity.

Misperceived Order

Related to the above, and perhaps a variant of it, is the human tendency to look for order in events. Order implies meaningfulness, and the search for meaning may be what distinguishes *Homo sapiens*. Certainly it is the basis of all art and science. (It is also, alas, the basis of all superstition.) Imagine rolling six dice, one after the other. The first one comes up with a 1, the second with a 2, the third with a 3, the fourth with a 4, the fifth with a 5, and the last with a 6. Most of us would remark on this, considering it highly improbable. If these dice were to be used in a high stakes game, one might be suspicious of the owner of the dice. Well, of course this outcome is in fact highly improbable: the probability of this is $1/6$th raised to the sixth power; it should only happen by chance 2 times out of a hundred thousand rolls! But assume the outcome was 4,5,3,3,2,4 or 1,1,5,3,2,4 or 2,5,3,4,6,1 or any of all the other possible outcomes, including the first six numbers of your home phone. The probability of *any* sequence is the same as any other—and incredibly small. What make us sit up and take notice when seeing a 1,2,3,4,5,6 outcome is that we recognize the coincidence that this is the sequence of the first six integers we learned at our mother's knee? There, of course, is nothing really special about it. However, creative individuals, being exceptionally aware and attuned to relationships, are going to see more order in random events than most folk. The dice come up 2,3,5,2,3,5 and the mathematician notes that these are the first three prime numbers repeated twice! (Or is amazed at a roll of 3,1,4,1,5,1 which are the first five digits of pi—ignoring the decimal place and with the last number being unimportant *really*, because a die doesn't have a 9 on any side. Or shocked at a roll of 1, 1, 1, 2, 3, 5 because these are the first six digits of the Fibonacci Series—if you accept that the first 1 in the series is the best the die could do given that a die doesn't have a zero side.)

Incomplete Understanding Of Probability

Most scientists, but few others except *smart* gamblers, have a grasp on probability theory. Even statisticians can sometimes be duped by some counter-intuitive results in this field.

The Gambler's Fallacy is the best known and most egregious example. If I were to flip a 'fair' coin (one known to come up heads half the time and tails the other half), and it came up heads twenty times in a row, I would intuitively think betting on tails next time would give me better odds. Tails are way overdue, right? Of course, this is nonsense. The odds of either heads or tails on the next flip are still equal. Coins don't have memories.

There are more subtle examples. One is the "Three Doors Problem", and another is the "Same Birthday Phenomenon".

The "Three Doors Problem" (also known as the "Let's Make A Deal Problem") involves the following. Imagine there are three doors, where behind one door is a pot of gold and behind each of other two doors stands a stinky goat with chronic halitosis. I claim psychic powers. I am asked to pick a door, and I do. Then you open one of the doors I did not pick and which also does not have the gold behind it, and ask me if I want to change from my original choice. We have agreed that if in the end I pick the door with the gold behind it, I get to keep the gold, *but* if I pick the door with a smelly goat behind it I have to pay you the equivalent value of the gold *plus* an extra five dollars! We also agree I have to play the game a hundred times to allow probability theory its just due. Of course you believe this is a sucker bet, for my chance of picking the door with gold pot behind has to be only 1 out of 3. There is no way those odds can change simply because you show me that one of the doors I didn't choose was a loser door. They certainly can't be better than 50/50.

I close my eyes and concentrate, mutter about astro-projection, and then decide to switch. (And I always decide to switch.) And you end up very poor, for my actual odds of getting the gold pot is 2 out of 3! And now you believe I must really be psychic and am probably banned from Las Vegas casinos.

I will leave it to the doubting reader to search out the explanation for this mysterious apparent changing of the odds. I admit that I was very sceptical when first presented with this problem, and so I went home and programmed the game on my computer to test it empirically. When it quickly became obvious that switching pushed my odds of selecting the gold up to $p=.67$, I then searched out the logical, theoretical reason for the advantage to switching doors—which I won't give away. To my admittedly nasty amusement, one of my colleagues, who taught advanced statistics, became extremely annoyed when presented with this problem and insisted adamantly that I was wrong. He looked a little sheepish the next day, but I had the tact not to mention it again.

"The Same Birthday Phenomenon" is something I always ham up when presenting it to my classes. I facetiously claim to be psychic and feel "vibes" about some astrological connection between two of my students in the front rows. "Born on the same day, bonded by stars!" "What," I ask, "is the likelihood of two people having the same birthday and sitting only seats apart in my lecture theatre? There are

365 days in the year, for heavens sake!" I challenge any sceptical student to bet $10 against my finding two students with same birthday from the first few rows. I always get a taker.* Then I ask them, one by one from the beginning of the first row, to say aloud the month and day of their birth, and ask anyone in the first few rows to stand up if it is also their birthday. Each row of the lecture hall seats at least twenty students. Statisticians know that—counter to 'common-sense'—the actual probability of finding two students having the same birthday is already in my favour by the time I just start the second row ($p=50.7\%$ for 23 students). My odds are up to 70% (the p for 30 students) by the time I've gone through half of the second row and 97% (the p for 50 students) if I have to go on into the third row—which happens very rarely. (Again, I leave it to the reader to hunt up the theoretical explanation for this.)

.

The point is that, like magic tricks, counter-intuitive probabilities often baffle perfectly intelligent people. If one is not trained in—or inclined to—scepticism, it is not surprising they can lead people down the garden path to absurd beliefs.

.

Projection.

Projection, as previously mentioned, is a Freudian term for the tendency to project our own concerns and interests and experiences onto ambiguous stimuli. It is the basis of psychological 'projective tests' such as the Rorschach ('Inkblot') Test. It is also the basis for 'psychic' con jobs and astrological readings.

.

Again I'll use as an example a prank I play on my Introductory Psychology students in the hope of instilling some healthy scepticism about things dressed up as science. I tell them that they can volunteer to participate in the final development of a new psychological personality test called the Barnum-Psi Personality Inventory. Their job will be to take the test and then 'validate' their resulting personality profiles for accuracy. I make a big deal about how it is going to be sent to a major research institute in Toronto for scoring, and how important it is that they use their own self-knowledge to fairly evaluate the personality profiles resulting from the scoring. The test consists of two parts: the first part is a true/false test which is a standard personality test called the EPI (the Eysenck Personality Inventory) and the second part is modeled on a projective test called the TAT (the Thematic Apperception Test) which involves writing

---

* I'm kind and never collect on my bet.

brief narrative stories based on interpreting ambiguous pictures of people interacting.* Their tests are never sent to Toronto, never even looked at by anyone! Instead the front ID sheet containing their pseudonym is attached to an "Individual Personality Profile" (which is identical for all participants) and returned to them for evaluation as to accuracy. The following is what they receive back from the analysis that never actually happened.

CONFIDENTIAL PERSONALITY SUMMARY AND INTERPRETATION

THE FOLLOWING EVALUATION OF YOUR PERSONALITY IS BASED ON A LENGTHY SCIENTIFIC ANALYSIS OF YOUR RESPONSES ON THE BARNUM PSI TEST. HOWEVER, FOR EASE OF INTERPRETATION YOUR PERSONALITY TRAITS ARE DESCRIBED BELOW IN NON-TECHNICAL TERMS. READ IT CAREFULLY AND THEN TURN TO THE NEXT PAGE AND RATE IT FOR ACCURACY.

You have a need for other people to like you and for them to admire you, and you also have a tendency to be critical of yourself. You have a great deal of unused capacity that you have not yet turned to your advantage. You are above average in creativity. While you do have some personality weaknesses, you are generally able to compensate for them. Your sexual adjustment has presented some problems for you, and during adolescence you had some conflict with your primary caregiver. But these issues are being resolved. While reasonably disciplined and controlled on the outside, you do tend to be worrisome and insecure inside, and at times you have serious doubts as to whether you have made the right decision or done the right thing. You prefer a certain amount of change and variety and become dissatisfied when hemmed in by restrictions and limitations. You pride yourself on being an independent thinker and do not accept others' opinions without satisfactory proof. You have found it unwise to be too frank in revealing yourself to others. At times you are affable, sociable and extroverted; while at other times you are reserved, wary and quite introverted. Some of

---

* I don't use the TAT images, but rather five pictures by the surrealist artist Rene Magritte.

> your aspirations may be somewhat unrealistic, but you have
> great determination to succeed. Overall you have a solid
> grasp on reality.

They are asked to rate this 'personal' description of their personality for accuracy on a Likert Scale from 1 (excellent) through 4 (average) to 7 (totally inaccurate). Every year the most frequently given rating is a 2 (very good) with 1 (excellent) or 3 (good) being the usual second most frequent choice. Only a handful out of 400 students last year gave it an average or below average rating for nailing down their personality.

This 'individual' personality profile is my personal variation on the original one I believe was developed by the psychologist Paul Meehlat which, in turn, was based on a newspaper astrological column. The Barnum in the test's name refers to P.T. Barnum, of circus and freak-show fame, who sagely observed that "there is a sucker born every minute." The "Psi" in the title gives it some scientific cachet.

After debriefing my students—and promising them that I will never again deceive them for pedagogical purposes*—I analyze how this evaluation, through the power of projection, can be so easily perceived as personal and accurate. It demonstrates why psychic and astrological readings are so often perceived as being incredibly accurate. There are four main strategies used in creating a convincing 'reading'.

- Be complimentary but not excessively and suspiciously so
- Keep it very general, so they can invent the specifics
- Make statements that are virtually true of everyone
- Use the 'shot gun' approach.

Using these strategies will make it easy for the unsuspecting dupe to project his or her own concerns and experiences into what is actually expressed..

Consider "While you do have some personality weaknesses, you are generally able to compensate for them." This could be paraphrased into "while you are not perfect, damn near!" The compliment is not

---

* Many of them don't believe me, but that's okay because although my intention was to make them skeptical even of what tries to pass for science, not skeptical of me, a skeptical attitude toward authorities is also good.

so obvious as to make one suspicious. It sounds critical but isn't really.

Consider "You prefer a certain amount of change and variety and become dissatisfied when hemmed in by restrictions and limitations." This is a nice general statement that doesn't really say anything except that you like a little variety and dislike being made to toe the line.

Consider "Your sexual adjustment has presented some problems for you, and during adolescence you had some conflict with your primary caregiver." Who, of course made it through adolescence—never mind adulthood—without some sexual problems? And who didn't have any conflict with their parents or guardians while growing up?

Consider "At times you are affable, sociable and extroverted; while at other times you are reserved, wary and quite introverted." Well, that about spans the spectrum, doesn't it?! Only the hermit and the truly extreme party animal could claim this is inaccurate.

In all these cases, one easily relates the vague and general statements to personal experiences and self-image. That reference to sexual adjustment presenting some problems easily becomes amazement that the psychic or astrologer or psychologist somehow knows about your rough time with Harry or Susan—or your first disastrous sexual experience.

One final thing worth noting is that undoubtedly some of the effectiveness of this Barnum Psi con is because it is dressed up in the trappings of science. I'm fairly confident that if I showed up in long flowing robes muttering about auras and planets in conjunction and then presented this reading after feeling the bumps on their heads, they would have read their 'personal' profiles a lot more sceptically and rated them as less accurate than they usually do. These are, after all, university students who presumably have been taught to be more suspicious about mumbo-jumbo than the less educated. But they have also been taught to respect science, and so they let their guard down.

False Memories

It may very well be that Freud's theory of the unconscious or subconscious is his most influential idea. This is a mixed blessing—to say the least. Certainly one of the most dubious and dangerous corollaries of this theory is the concept of repression. According to traditional Freudian theory, traumatic or disturbing events are repressed; i.e., driven from conscious memory and exiled to the realm of The Unconscious. Much of Freudian psychoanalysis is devoted to bringing these allegedly repressed memories back to consciousness by free association or dream analysis.

The actual evidence for repressed memories is almost non-existent, except (maybe) for so-called fugue states and drug-induced amnesia. In fact, normally the opposite of repression is true: traumatic events are remembered only too well and vividly—as happens in Post-traumatic Stress Disorder. It is not a topic for this book, but the damage done by psychologists claiming to "uncover" repressed memories is well documented.* They implant memories, rather than uncover them.

What *is* true is that memory is not to be trusted. The research on this is unambiguous and there for anyone to search out, but we still have great and unfounded faith in the accuracy of our memories—especially what is called *episodic memory*. There are several different kinds of memory, and recent discoveries using brain scanning technology have even localized these different types of memory. We have memories that are sensori-motor, such as remembering how to ride a bike. We have memories of information, usually called *declarative memory*. We have memories of life events, usually called *episodic memory*.

Although I've already considered the nature of memory in the broader context of its role in creativity, it is worth reiterating and emphasizing what is known about memory, for while its fallibility may serve in creative endeavours, it can also be the reason for the emotionally-charged obstinacy with which people maintain beliefs in experiences that simply didn't or even couldn't have occurred. If it is

---

* For more information on this, read *The Myth of Repressed Memory* (1994) by Elizabeth Loftus. She is Distinguished Professor at the University of California, Irvine, and has conducted numerous experimental studies on the implantation of false memories. As mentioned in the earlier chapter on memory, pointing out to someone that their episodic memory is inaccurate often makes people very angry. Dr. Loftus received death threats and required security protection at conferences for pointed out the simple fact of memory's fallibility and the very real dangers to human rights excessive faith in memory presents.

hard to convince people that they can't always trust their senses, it is often even harder to convince them they can't always trust their memories.

.

We usually realize the unreliability of declarative memory. Students demonstrate this when they write exams and stay till the last possible minute trying to decide whether the right answer on a multiple choice question is alternative B or C. Most of us when questioned about facts realize we might remember something incorrectly, and we aren't usually emotionally committed to what we think we remember. To stand firm on every fact we think we remember is just asking to be humiliated, for someone is bound to present concrete evidence to correct us when we are wrong.

.

It is a totally different story with episodic memory. For some reason we passionately believe that our memories of past events are infallible. Many a domestic dispute results from this ill-founded faith in our episodic memories. Episodic memories are crucial to witness testimony in the courts, and good lawyers have always been expert at demonstrating their unreliability. Recent research has shown that episodic memories, especially of traumatic or violent or highly emotional events, are so unreliable that some philosophers of jurisprudence are arguing that the legal system's reliance on witness testimony needs to be completely overhauled.

.

There are numerous places where memory can be corrupted. A standard schema to describe memory is to divide it into three parts: perceptual input; storage; and retrieval. Things can go wrong at every stage. We can misperceive. We can damage or misfile the information. We can fail to find the information—or find the wrong information—during the retrieval process. However, I think the major problem with accurate memory of events is best explained by a simple metaphor.

.

I want to remember something, so I go to the filing system that is my brain. I hunt around and pull out a copy of the file. I read it. *And then I edit it!* When I'm done with it, I replace the old version with the edited version. Every time I pull that file I make more 'corrections'— often in line with my current beliefs and prejudices. Paradoxically this means that the most often recalled memories may be the least accurate. My wife constantly remarks on how the anecdotes I repeatedly drag out at parties keep getting more elaborate.*

With declarative memory items, this is less likely to happen, for facts are usually checked when the file comes out. However, usually there is no fact checker around for recalled memories of events. Compounding the problem is that the memory system that stores episodic memories is linked to the emotional system of our brains—the limbic system. So not only do we not get called on any errors—which is likely to happen with declarative memory—but we invest them with emotion and defend them tooth and nail.

The usual evidence cited for paranormal experience is anecdotal, episodic memories. My aunt tells me she remembers clearly waking up at precisely 3:12 a.m. the night her father died, and I have the tact not to challenge her. What probably happened is that she was worried about her father (who was terminally ill) and so she did sleep poorly that night, waking up repeatedly. The next morning she finds out that her father had passed on during the night and she makes the connection with her sleeplessness. Each time she retrieves and recounts the story she embellishes it before re-filing it, until eventually she believes she clearly 'remembers' looking at her bedroom clock and seeing the time—which matched perfectly her dad's official time of death.

It follows that an artist, whose major wellspring of inspiration is often episodic memory, and whose imagination is exceptionally powerful, is even more likely than most to think they have personal experience of paranormal phenomenon based on repeated visiting of past experiences. On the other hand scientists are more dependent on declarative memory and so less easily fooled.

Desire To Believe

Demosthenes had it right when he said we tend to believe something as true if we wish it to be so. Surely we all wish that our loved ones who have shuffled off this mortal coil could make contact with us from the Great Beyond. Beside the comfort of still being in touch with them, it gives the more selfish and reassuring knowledge that there actually is a 'Great Beyond'. Most (albeit not all), beliefs about the paranormal or the supernatural or the various species of pseudo-

---

* I like to think they get more interesting as well.

science are obviously about things we wish were true. It would be nice if we didn't just fade from existence at death. It would be nice if we could predict the potential future we would have if we made one decision as opposed to another. It would be nice if we could cure cancer by prayer—or by eating peach pits in some Mexican 'wellness' resort. Of course there are dark sides to many superstitious beliefs: heaven in many religions is accompanied by a hell; prophecies are sometimes dark and considered inevitable; belief in miracle or 'alternative medicine' cures is often paired with paranoid fears of poisons in our food or our traditional doctors' malevolence. Nevertheless, most—or at least many—of our dubious beliefs are rooted in wanting to believe, because it would be a lot nicer if these beliefs were true.

.

Again, perhaps the creative individual is often more "open to experience" (to use the psychometric term) and also more open-minded, which is to say more willing to consider alternative possibilities. Naturally the scientist tempers this with the scepticism that is central to his endeavour, but the artist has had less need for this scepticism in what he does.

.

.

I should conclude by saying that it is entirely possible that all these factors that predispose otherwise exceptionally intelligent people to irrational beliefs operate effectively on *all* people, eminently creative or not. Scientists just may be less susceptible than others because scepticism is so much an integral part of their creativity. Nevertheless, it remains disconcerting how many brilliant artists have subscribed to the most outlandish beliefs.

A major assumption underlying this book is that the rift between the arts and the sciences is healing because of increasing understanding—and utilization—of the skills and tools and philosophies of the two camps. Even given the reasons just suggested for artists' continuing susceptibility to superstitious beliefs, it nonetheless may seem that this continuing irrationality in the artistic community is evidence that a real coming together of artists and scientists may just be wishful thinking.

.

So, I think another disturbing trend has to be considered. Many concerned and informed social commentators have written about, warned about, the increased acceptance of religious and superstitious beliefs as having an inalienable right to "respect". Paradoxically and ironically, such well-meaning desire for fairness has resulted in the appalling political correctness and cultural relativism that has been amply satirized—albeit to little effect.

.

Good and fair journalism is expected to strive for 'objective' and 'balanced' reporting. The facile solution to this is to "present both sides of the story" without editorial comment or evaluation. This has resulted in the most absurd 'balanced' reportage: creationism versus evolution; parapsychology versus physics; tribal vengeance traditions versus common law. Consider the creationism versus evolution debate, which in the United States is actually taken seriously. Consider the reporter, who in striving to be 'fair' and diligent, searches far and wide to find someone with a science degree to stand on one scale of the balance, while knowing any one of 99.9% of scientists could stand on the other scale—but only one can be placed there. The impression, naturally, is of 'balance', resulting in both sides appearing to have equal credibility. And human nature is such that the more emotionally reassuring viewpoint is going to win out. Hell, if even scientists can't agree, go with what feels good!

.

Recently I've come across numerous 'balanced' articles about the flakey, mendacious, and malicious scientology cult, which counts among its 'congregation' a few talented actors with celebrity status. That the scientology con artists are given any credibility at all is not 'fair' journalism: it is irresponsible journalism.

.

As Richard Dawkins has most pointedly pointed out, there is no reason to afford 'respect' to crazy ideas or their proponents. But if someone in the name of religion mouths those ideas, then too often too polite (or intimidated) journalists, liberals, and just plain well-meaning folk hold their tongues. A scientist presenting his design for a perpetual motion machine at a conference would not be afforded any such respect. Nor would an artist presenting his Grade Three crayon scribbles to the jury for an art exhibition be afforded any respect. No doubt there would be much mocking and satirical remarks made in both cases. But satirizing religious figures or beliefs is asking for trouble—even violent trouble. The infamous Danish Mohammed cartoon episode is but one example.

.

The atmosphere of 'tolerance' and 'respect' for the most absurd, and often dangerous or destructive, belief creates an intellectual environment where every idea is considered, in good egalitarian fashion, equal. Of course, beliefs and ideas aren't really equal—just as people aren't equal. (As with people, each should have equal opportunity to compete, but that is a different thing entirely.)

.

Many are the doomsayers predicting a new Dark Ages. Certainly it isn't hard to find evidence to support such fears. Although the United States allegedly has a literacy rate close to 100%, the majority of Americans don't actually read anything and are notorious for their lunatic religious beliefs—such as the so-called "Rapture". That ill-educated folk in Islamic countries can be led to believe in crazy and dangerous notions such as "Martyrs' Heaven" is understandable, but when the supposedly educated citizens of the Western industrialized nations hold equally insane beliefs, one has reason to be concerned. As I write, three of the Republican hopefuls for the forthcoming American presidential have publicly stated that they do not "believe" in evolution! A properly credentialed fellow who taught at my university expounded the mystical power of pyramids and crystals. But one comes to expect ignorance in most politicians and, alas, too many academics. So it wouldn't be so bad if it weren't for the fact that some of the finest creative minds of our time (especially in the arts) have been infected with belief systems that rival those of the original Dark Ages.

.

However, as always a little historical perspective can temper our fears. Some of the most brilliant creative geniuses of the past embraced strange and irrational beliefs. Previously mentioned Newton's religiosity is but one startling example in the scientific domain. The examples in the artistic domain are legion.

.

My personal view is optimistic. As artists, the more easily infected by nonsensical ideas, turn more and more to science, they inevitably must come to absorb some of the healthy scepticism of their complementary discipline, especially as scientific knowledge increasingly becomes a very valuable—even essential—tool for use in their own endeavours. As for the general populace, even here I have some optimism. Accessibility to the world's accumulated knowledge is but a computer and an Internet Service Provider away. That the lunatic fringe has equal opportunity to promote its ideas on the Internet does not concern me as much as it does some. Most who believe in freedom of expression have an implicit belief that "truth outs".

.

However, the importance of computers and the Internet is a topic for the final section of this book about the future. Suffice it to say here that while poets may not really be (as Keats claimed) "the unacknowledged legislators of the world", artists and scientists really are those who chose the path humanity eventually follows, no matter how much resistance their directions and guidance initially encounters. Artists and scientists led humanity out of the dark woods of The Dark Ages during the age aptly named "The Enlightenment". So one kind of 'faith' I do have is that of the optimist's faith that we can be dissuaded from re-entering the dank, dark forest of ignorance—again by the collaborative efforts of artists and scientists.

CASE STUDIES: JAMES RANDI, ALFRED RUSSEL WALLACE

James Randi is a brilliant artist and Alfred Russel Wallace was a brilliant scientist. What makes them an interesting contrast is that Randi is the one possessing the virtue of scepticism, while Wallace is the one with what can only be called gullibility.

.

Magic is an art, and like much art is based on deception. Fiction is not true, even if it leads us to true insight. A super-realist painting or *trompe-l'œil* is not a three-dimensional object, even if it charms us by fooling our eyes. To say that magic is *just* trickery may be literally true, but that 'just' seems unjust, for so are novels and paintings just trickery. To be cleverly tricked is one of the pleasures of art. We only resent trickery when we don't know that it is trickery. The best selling novels of Carlos Castenada were passed off as non-fiction, as true stories of his anthropological research, and when he was exposed, this angered many who had been duped.* They remain good reading, but it is a different kind of read if you believe you are reading literal truth or know you are reading a convincingly contrived fictional narrative. The vast majority of magicians, like the vast majority of novelists, admit they are in the deception trade and actually derive their pride from admitting this and still being convincing.

.

James Randi (Randall James Hamilton Zwinge) was born in Toronto, Canada, on August 7, 1928. Like many kids he was fascinated by magic tricks. When a serious biking accident put him in a body cast for over a year, he devoted this time to getting very, very good at these tricks. He grew up to become a very successful stage magician performing under the moniker The Amazing Randi. His reputation grew when he followed in Harry Houdini's footsteps by doing apparently superhuman feats of endurance of escape. He even broke Houdini's record for being in a sealed casket: surviving for an hour and forty-four minutes, thirteen minutes longer than Houdini had. Randi also survived being frozen in a block of ice for fifty-five minutes. His 'escape' escapades were just as impressive, including

---

* His first three books (*The Teachings of Don Juan: A Yaqui Way of Knowledge; A Separate Reality; Journey to Ixtlan*) about his experiences apprenticing with an Indian shaman somewhere in Mexico were bestsellers and earned him a Ph.D. in anthropology from his gullible academic advisory committee at UCLA. I remember my scepticism about the veracity of the books annoying many of my contemporaries in those infamous Sixties.

escaping from straitjacket while hung upside down over Niagara Falls.

But Randi's real fame came when he set out to debunk the hucksters and con artists who claimed their magic wasn't just tricks, but 'real magic' or paranormal powers. As honest magicians confided in him their techniques and Randi then saw evangelists and con-artists using these tricks to convince people of their special powers—for their own personal gain, Randi began to spend more and more time and energy on debunking pseudo-science. Certainly this is very much to his credit as a highly moral person who just doesn't want to see people being bilked, but I'm sure there is a small part of him that is motivated by hating to see second-rate artists getting so much attention just by lying about what they are actually doing. (A great novelist would be annoyed if a second-rate hack got more attention, just because he pretended he was writing non-fiction.)

Most notable of his targets was Uri Geller, who back in the early seventies was making fools out of everyone, including scientists at Stanford and writers for the popular magazine *Psychology Today**, and getting a lot of credulous publicity. Geller is a very a litigious guy, quick to sue anyone who denies his supernatural powers for outrageous sums, and when Randi pulled no punches in exposing him, Randi was immediately embroiled in a high-stakes lawsuit initiated by Geller. Geller was noted for the cheap conjuring trick of allegedly bending spoons with his mental powers, but he wasn't able to bend the courts' mind, and he hasn't won any of his numerous suits.

It is ironic that Randi's fame was to some extent created by the established fame of an individual more deserving of infamy than fame. No matter. Randi has used his celebrity to advantage—everyone's advantage. With public appearances and his writings he certainly has done more to educate people about pseudo-science and the dangers of gullibility than those in our academic institutions charged with that responsibility. In 1996, Randi established the James Randi Educational Foundation (JREF) devoted to debunking pseudo-science. He has set up a trust fund that will award one million

---

* See the 1974 June and July issue of *Psychology Today*. Geller's alleged powers were the subject of a study by Puthoff and Trag at the Stanford Research Institute where they allegedly concluded he has sufficiently demonstrated 'psychic' powers as to warrant further study. Unfortunately for Geller, these two 'scientists' are noted for flakey endorsements of all kinds of strange ideas, including 'remote viewing.'

dollars to anyone who can demonstrate under controlled conditions any of the paranormal phenomena that so many naively believe in. Needless to say, the fund is collecting lots of interest, and no one has collected the million. The ultimate, very scientific test is "Put your money where you mouth is." Randi has. The failure of any person with alleged supernatural or paranormal powers to collect on this 'easy' million is more convincing, real world evidence of the bogus nature of parapsychology than any number of well-reasoned arguments.

.

It is to be expected that while he is highly respected by scientists, who have to appreciate him doing the dirty work of debunking pseudo-science, many, many people also hate him. If you believe in some nonsense, you're not going to like the person who keeps pointing out that it is nonsense and makes fun of your belief. And if you're the one conning, not the one being conned, you are going to be even more hostile to this guy giving away your tricks.

.

Randi, like most creative individuals, can be very disagreeable, outspoken and abrasive. However, he doesn't attack self-deluded individuals, although he will try to free them from their delusions. His often-vitriolic wit is primarily directed at those he knows, or suspects to be, shamelessly cashing in on other people's gullibility, and causing harm in the process. Many so-called "alternative medicines" and "psychosurgery" are particularly egregious examples of doing real harm—by luring people away from proper medical treatment. When Randi was awarded a MacArthur Foundation "Genius Award", he used the money to expose the dangerous fraudulent 'faith' healers Peter Popoff, W. V. Grant, and Ernest Angley.

.

An important lesson to be learned from Randi's example is that artists can be more scientific in their thinking than scientists. Which brings us to Alfred Russel Wallace and another irony.

.

Wallace was born on January 8, 1823. A man of eclectic interests and a bit of an explorer: he traveled widely, including the Amazon River basin of Brazil and the Malay Archipelago between Australia and Indochina. To some extent he emulated Darwin's famous voyage on The Beagle with his own voyage on The Mischief. It is fair to say he was a naturalist, a geographer, an anthropologist, an entomologist, a social reformer, and an ecologist—at a time when these disciplines weren't professionalized. Like Galton, he was a bit of a post-Renaissance Renaissance man. Also it is fair to say, if unkind, that in the later years of his life, he was duped by pseudoscience and expressed belief in numerous dubious ideas that now seem a sign of the credulity one finds more often in artists than scientists.

Of course Wallace is best known as the co-discoverer of The Theory of Evolution By Natural Selection. But often that is all that is known about this exceptional man. Many feel it is unfair that Darwin gets all the glory.* Darwin was delaying publication of his findings and his over-arching theory when he was persuaded by friends to get on with it, because Wallace's numerous publications were going to steal his thunder. Probably if Darwin had not been convinced to go ahead with publication with *The Origin Of Species*, the Creationists would be attacking Wallacism, not Darwinism. And Wallace's latter descent into spiritualism would have supplied them with more ammunition than the myth of Darwin embracing religion on his deathbed.

What isn't well known is that Wallace and Darwin corresponded extensively and shared ideas, and collegially disagreed about many aspects of their respective theories regarding the mechanism of evolution. What also isn't well known, and especially among those silly folk going on about "evolution is just a theory" is that the fundamental idea of evolution was around long before Darwin or Wallace—albeit this "transmutation of species" wasn't yet entirely endorsed by the current scientific community simply because it seemed inexplicable by natural causes. So it was the *mechanism* by which it occurred that was really mysterious and controversial, and it was the Wallace-Darwin explanation called *natural selection* that was truly revolutionary and responsible for one of the greatest paradigm shifts in the history of science.

There are several excellent books that deal with Wallace in the context of the development of evolutionary theory†, including much scholarship about his relationship with Darwin, including their differences in their understanding of the most important mechanisms of natural selection. Overall, however, they were staunch supporters of each other even when they disagreed about details. Papers by both of them about the seminal idea of natural selection were presented jointly at the Linnaean Society of London in 1858—through Darwin's arrangement. Their mutual admiration is obvious by

---

* Lately, his importance is being rediscovered and appreciated.
† These include works by Loren Eisely and Daniel Dennett. The frequently recommended biography is *Alfred Russel Wallace: A Life* by Peter Raby. There is also a scholarly website devoted to Wallace that is easy to find.

constant reference to each other in their publications on evolution. And Darwin, who had greater stature and influence in the current scientific community and was aware of Wallace's continuing financial difficulties, even paid him for some editing work and lobbied the government to give his colleague a pension for his contributions to many areas of science.

.

But what is of relevance here is not this interesting relationship of two geniuses or the importance of evolutionary theory, but rather the nature of Wallace's creative personality that is so contrary to the stereotypical view of the conservative, hard-headed, and extremely sceptical scientist. He *was* a hard-headed, sceptical scientist in many ways, but he also had an attraction to radical ideas that some would call open-mindedness and some would call naiveté.

.

As the hard-headed scientist, he designed a simple empirical proof of the curvature of the earth in response to a wager from the flat-earth nutcase John Hampden. (This earned him a much-needed £500, but the litigious Hampden, through constant lawsuits, cost Wallace more in court costs than the original prize money.) * He wrote a whole book (*Is Mars Habitable?*) to refute Percival Lowell's claims about the existence of Martian canals. And of course, he is the co-discoverer of natural selection and staunch defender of it in a time when it was still controversial.

.

On the other hand he waged a war against smallpox vaccination, convinced that it didn't work, that the germ theory of disease was suspect, and that better hygiene was all that was required. And then most distressingly (and most damaging to his reputation), he embraced spiritualism and defended séances as empirical evidence of the reality of an afterlife and communication with the dead. Although he continued to reject traditional religion and maintained that he was a scientist, first and foremost, he eventually came to believe that at certain points in evolution some God stepped in and did a little magic to keep things moving along: 1) at the initial creation of life; 2) at the formation of consciousness; and 3) at the beginning of the species *Homo sapiens*. Furthermore, falling victim to the teleological urge, he came to believe human beings were the ultimate purpose of evolution—in some sense making him an early "Intelligent Design" advocate. In 1903, this last belief was brilliantly satirized by Mark Twain (who didn't seem to consider our species as exactly the most

---

* Does this sound familiar? Remind anyone of Uri Geller's response to Randi's debunking?

noble of all creations) in a little known piece entitled "Was The World Made For Man".*

What to make of this apparent inconsistency? Biographers all seem to agree that Wallace had no respect for authority, was naturally attracted to radical ideas, and was quite willing to investigate phenomena or hypotheses that were considered 'flakey' by the establishment. These are admirable traits and traits common to the creative individual. He shared with William James a belief that "there is no source of deception in the investigation of nature which can compare with a fixed belief that certain kinds of phenomena are impossible."†

Wallace was fascinated by phrenology (the assumed correlation of bumps on the head with personality traits) and mesmerism (hypnosis)—which now we view with justifiable scepticism. But phrenology was the primitive precursor to neurology, as astrology was to astronomy, and alchemy to chemistry. Things have to be put in historical perspective. A few scientists even now take hypnosis seriously, even though it was originally based on some assumed magical properties of magnetism. Human suggestibility is a real phenomenon, and no one is immune. Even Wallace's belief in spiritualism is not that strange, for in the Victorian Era it was not nearly as absurd as it is today. Otherwise famously rational people, including Conan Doyle and the great scientist Crookes thought it 'scientific'. It is wrong and a marker of ignorance of history to blame scientists for being wrong or credulous by our current standards if their open-mindedness sometimes led them to false conclusions.

And what is the moral of the story of these two exceptional and admirable individuals—Randi and Wallace? There are several. Scepticism and rationality is not only the purview of the scientist. Although these traits are more associated with scientific creativity, creative scientists are also susceptible to the blandishments of teleological thinking and sometimes overextend their scepticism (as with Wallace and the effectiveness of inoculation) and sometimes aren't sceptical enough (as with Wallace and spiritualism). And artists, such as James Randi, can sometimes be more scientific, more rational, more effectively sceptical and sensible than professional

---

* This very humorous essay written for *The Scientific American* can be found on the Internet and is well worth reading. Twain had a scientific bent.
† Often quoted, this from a letter in 1886 to his friend Carl Stumpf.

scientists. (Geller repeatedly fooled trained scientists, but Randi saw right through the con.)

.

The gift of creativity—and its attendant traits of open-mindedness and scepticism—is a double-edged sword. Not all those so gifted only cut through the bullshit; sometimes they cut themselves or flail at imagined dragons.

.

We don't need Him to explain things, but what the hell: God bless!

EPILOGUE: IMPROVING OUR VISION

"The most pathetic person in the world is some one who has sight but no vision."
—Helen Keller

"There are poisons that blind you, and poisons that open your eyes."
—August Strindberg

Creativity is a mixed blessing. This book may seem to view it through dark glasses. But if our glasses are rose-tinted, we can't see things as they really are.

.

When you realize you can't see clearly, you should purchase new glasses. Our understanding of creativity is still far from clear-sighted. It may be that we need new glasses. The world may at first seem darker, but clarity is always to be valued.

.

Science and art can help diagnose our weaknesses, and offer ways to compensate for them. However, even when the evidence is strong that you aren't seeing things as they really are, it is tempting to deny it. Science and art can help diagnose our weaknesses and offer ways to compensate for them. But they cannot force us to deal with them. We need to have the will and courage to do so.

AFTERWORDS

In other words, stuff stuffed at the end.

ACKNOWLEDGEMENTS

I think virtually everyone, upon beginning to read a book, skips the acknowledgements—everyone, that is, except those who feel they should be acknowledged. Better, it seems to me, is to do as they do in the movies: put the credits at the end.

I'm limiting my acknowledgements so those who have been so particularly helpful and supportive they shouldn't be buried in a huge list that no one will read. However, I also would like to extend an all-purpose, less specific, thank you to some other folk as well.

There is no way I adequately thank my wife, Ursula, for all she has done for me. She has always been an indefatigable editor. Without her, I would've been continuously embarrassing myself in print from the very start. But most importantly, she has been the inspiration for all my work. She redeems my life. She is my Beatrice—and I got to actually marry her and not just admire her from a distance.

I also owe a debt of gratitude to our two 'kids', Christiaan and Katherine. They certainly deserve credit for their tolerance of their eccentric father, but, more importantly, for their passion for all the aesthetic and intellectual, and even purely physical pleasures, life has to offer. They have are an inspiration to me.

My family has taught me that loving another and nurturing that 'selfish gene' is the most creative thing we can do. And the most rewarding.

I also am incredibly grateful to the many scientists and artists whose creations were the ultimate inspiration for this book—and even for my way of life.

Thanks, too, to those several teachers I had long ago who through their own passion for art and science instilled in me the respect I have for creative endeavour, as well as to my students who have had to listen to me test out my ideas on them in what I'm sure were sometimes rather rambling lectures in my Psychology of Art class.

I should also acknowledge Nipissing University for giving this writer a very rewarding day job for 40 years. And much of the first draft of *Secret Agents* was written during a sabbatical leave.

SELECTED ANNOTATED BIBLIOGRAPHY

In my judgement the following works are well worth reading (or at least sampling) and *not* because I think they offer supporting evidence for my various theses. (Some do not.) I recommend them because I believe they are all entertaining and enlightening works that are at least tangentially related to the content of this book. Some are about creativity and creative individuals and some are the actual works of the creative individuals mentioned in *Secret Agents Present*.

.

Recommending books is way too much fun, for one loves to plug books and writers one admires, but I've tried not to over-indulge in this pleasure. Life is very short—even if one reads quickly—and so I've grouped my recommendations by the sections, so that if something along the reader's way triggers a particular interest, he or she can easily follow up on what inspired me. I've almost always limited my recommendations to a maximum of six per chapter, usually with two of these referring to the subjects of the 'case studies' for that chapter. The list is still long, but rest assured—there will *not* be a quiz!

.

I admit that what I have chosen to include is extremely arbitrary and idiosyncratic and reflects my own reading rather than some omniscient overview of the many topics touched on. Anyone intimate with a particular chapter's theme could surely cite more relevant and better works, even though I've sincerely tried to select, given my own limited knowledge, excellent works I believe touch most of the bases of each section. I've included a very brief justification for the inclusion of each work in this list.

.

I've usually only given the author and title of the work, for in The Amazonian Internet Age that is sufficient. The only exceptions to this are when more information is really required to easily find my recommended material.

- **Through A Glass Darkly**
  - General suggestions and remarks.
    - o Everyone is familiar with the ideas usually attributed to Locke, Rousseau and Descartes, for they are ideas that are accepted as unthinkingly now by the majority of 'right-thinking' people as was God's goodness and Man's centrality in the universe five hundred years ago. We are infinitely malleable (Locke), innately noble but corrupted by civilization (Rousseau), and still unique, somehow separate, from the mere material world (Descartes). Unfortunately for our egos and sense of empowerment, the hard evidence fails to support any of these contentions. Examination of the nature of creativity requires evidence-based reasoning—not wishful thinking. So my one recommendation for the introductory chapter to this section is by a thinker who epitomizes this tough approach.
  - Recommended reading.
    - o Pinker, Stephen. *The Blank Slate: The Modern Denial of Human Nature* – Pinker is usually called a "cognitive scientist"—a label that doesn't begin to describe the man. Certainly he is an eminent scientist who has made major contributions to understanding cognition, especially the study of language, but he also is a philosopher of the first order. Furthermore, he belongs in that elite pantheon of thinkers who have the *cojones* to tell the ruling emperors of putatively liberal thought that they are running around starkers. I cannot recommend this book too highly, for its many virtues include honesty, wit, insight—and compassion grounded in reason.

- **Axiology And Epistemology: The Amorality Of Creativity**
  - General suggestions and remarks.
    - o It isn't difficult to find evidence of the amorality and even immorality of creative persons. Any honest, or unauthorized, biography of almost any major creative individual will do that. Or one can consult memoirs from former lovers; for example, Françoise Gilot's *Life with Picasso*. Even writers occasionally pen autobiographies that are brutally honest about their own misbehaviour; for example, Tennessee

202

Williams' *Memoirs*. The best of these uncover the unfortunate connection between the person's creativity and his or her less than admirable behaviour toward others. There is a perverse pleasure in reading these works, but it is worth remembering the injunction that those in glasshouses shouldn't throw stones.

- Recommended reading.
    - o Burgess, Anthony. *A Clockwork Orange* – This novel published in 1962 and made into an infamous and brilliant film by Stanley Kubrick in 1971 has among its many themes the frightening application of psychological theory to social engineering. However, I choose to include it here because of its theme of the independence of ethics from aesthetics. The morally brutal young man who is the main character is also a sensitive aesthete who adores Beethoven. In the film, the juxtaposition of heart-breakingly beautiful music with just as beautifully choreographed but appalling violence somehow makes this violence more deeply disturbing than that even the most graphic of 'slasher' movies or even the works of that cinematic master of violence, Sam Peckinpah.
    - o Miller, Henry. *Colossus of Maroussi* and *Tropic of Cancer* – The totally charming *Colossus* is a corrective to the common misconception that Miller is nothing more than a crass writer of pornography. On the other hand, *The Tropic of Cancer* is a modern masterpiece and a shameless—and fearless—self-exposure of artists' amorality when obsessed with what they want to create.
    - o Nietzsche, Fredrich. *Ecce Homo* and *Beyond Good and Evil* (the Kaufman translations) – Nietzsche was one of the first philosophers to argue against the conflation of art with ethics, although he did see what he considered decadent art as a symptom of a civilization in decline. Even more importantly his famous conception of the *Ubermensch* was of an artist who was "beyond good and evil".
    - o Shapiro, Karl. *In Defense Of Ignorance* – I include this book partially because it contains one of the best critical essays on Henry Miller I have ever come

across and partially because it—well, because it is just great stimulating reading by a fine poet.

- o Shelley, Mary. *Frankenstein* – This is still great reading and a work that both represents the fear of science among the Romantics while simultaneously showing the potential for empathy we should have with any 'monsters' science may produce.

- **Madness And Genius: Allies Or Enemies?**
  - General suggestions and remarks.
    - o As with interesting reading about amorality among the creative, so too is it easy to find juicy bits about insanity in the same sources: biography, autobiography, and former friends' and lovers' memoirs. So I've merely included two quasi-scientific books on the topic and biographies—those of Ada Byron and her mad, bad Dad.
  - Recommended reading.
    - o Jamison, Kay. *Touched with Fire: Manic-Depressive Illness and the Artistic Temperament* – Jamison may have biases because of her own bipolar disorder and her psychiatric training, but her defence of the thesis that this heritable chemical imbalance in the brain is linked to creativity is convincing.
    - o Ludwig, Arnold M. *The Price of Greatness: Resolving the Creativity and Madness Controversy* – No one should talk about the relationship of madness to creativity without reviewing Ludwig's research. I'm not sure he "resolves" the conflict, but he certainly presents some interesting data.
    - o MacCarthy, Fiona. *Byron: Life and Legend* – This huge book is extremely comprehensive, but it is detail that brings a biographical subject to life. The devil (and the fun) is in the details.
    - o Wooley, Benjamin. *The Bride of Science: Romance, Reason, and Byron's Daughter* – I felt I that it would be unbalanced to not include a pointer to a book about Ada Byron since she is the foil to her father in the case studies for this chapter. However, I have to confess I have not read this book (only a brief excerpt and a few mixed reviews), so I'm including it with that caveat. Unfortunately, few books deal with this important woman, and because this one

supposedly considers the conflict between her artistic and scientific nature as well as her emotional instability, I have included it.

- **Our Sensory Censors: Our All Too Human Limits**
  - General suggestions and remarks.
    - o One branch of psychology that is unquestionably scientific (largely because it is really biology and neuroscience) and highly relevant to the artistic endeavour is what is called sensation and perception. Sensation refers to the physical mechanisms that transduce information from the external world (be it electromagnetic energy, pressure waves in air, some chemical invading our mouths or noses, etc.) into nerve impulses carried to our brains. Perception has to do with what our brains make of these nerve impulses: our experience of this transformed information. This is a fascinating subject of which artists have always had—and needed—a working knowledge, and scientists have now made major inroads to understanding. Recommended is almost *any* standard textbook (because I couldn't possibly review and evaluate the many in print) with the predictable title of *Sensation and Perception*. The study of this subject is a humbling corrective to the assumption that our senses can be trusted to accurately report to us on the goings-on in the 'real' world.
  - Recommended reading.
    - o Baron-Cohen, Simon and Harrison, John E. (editors). *Synaesthesia: Classic and Contemporary Readings* – As the title indicates this is a collection of writings about the mysterious and fascinating phenomenon of synaesthesia.
    - o Hickey, Thomas J. *History of Twentieth-Century Philosophy of Science* – This book is still available for purchase as a paperback, but the author has generously posted a version on the Internet where it can be read or downloaded for free in PDF format. It is big, and it is comprehensive, but it is certainly worth diving into for its insights into how many of the methods and assumptions of scientists are as diverse, dubious and biased as those of artists.

(http://www.philsci.com/index.html)

- o Hawking, Stephen. *A Brief History Of Time: From the Big Bang to Black Holes* – It seemed surprising to many, but it doesn't seem strange to me, that this book also made the best-seller list, for it is well-written, easy reading, and deals with some of the most mind-boggling scientific ideas of all time (pun intended). It also gives the reader a glimpse into the amazing mind of Stephen Hawking.

- o Levitin, Daniel J. *This Is Your Brain On Music* – This very readable book is the best book on the nature of music, its creation and its appreciation, I have come across. As a musician turned neuroscientist, Levitin is uniquely qualified to 'explain' the magic of music. That he also writes with such clarity and wit explains why this book has become both a best seller and received rave reviews from both musicians and scientists.

- o Seckel, Al. *Incredible Visual Illusions: You Won't Believe Your Eyes* – Despite its tacky title, this book is quite informative as well as entertaining. (And of course it should be said that convincing people to not believe their eyes so easily *is* good advice.) It contains hundreds of visual illusions organized by type with explanations of how they might work. It is fun to look at and informative to read. The author is a cognitive vision scientist at CALTECH and certainly knows what he is talking about.

- o Solomon, Maynard. *Beethoven* – This biography is considered definitive by most historians and musicologists. As with Mozart, the literary or filmic portrait is more caricature than accurate portrait. Good biographies, such as this, remove none of the tortuous drama, but offer real insight into their subjects.

- **Memory, Language, And Thought: Our Unmetrical Mental Maps**
  - General suggestions and remarks.
    - o It is absurd how many psychology textbooks dealing with the topics briefly addressed in this chapter are published, especially since these books are virtually interchangeable. (The creation of textbooks, which

have a guaranteed market if instructors can be lured into using them, is a major industry.) Still, the topic *is* an important one. Probably the best quick and concise overview can be obtained by reading the relevant chapters (sure to be present) in any fairly recent Introductory Psychology text. However, I've included in my idiosyncratic list two books of far higher literary quality and depth that deal specifically with language (by Pinker) and memory (by Schacter).

- Recommended reading.
  - De Botton, Alain. *How Proust Can Change Your Life* – This is a book I've included solely on the basis of reviews, the reading of a small excerpt, and an intuitive feeling that it says much about the themes so dear to my heart.
  - Pinker, Stephen. *The Language Instinct: How the Mind Creates Language* – This a lucid and entertaining review (by a leading expert in the field) of what we know about language acquisition and how our brains manage to convert those pressure waves hitting our eardrums into something as profound as our intellectual and emotional and aesthetic reaction to a Shakespearean sonnet. Not every language expert will agree with all of Pinker's conclusions, but he is scrupulously fair in describing alternative hypotheses and always presents cogent arguments. He also obviously derives a lot of fun with language, which he shares with his readers.
  - Rucker, Rudy. *Infinity and the Mind: The Science and Philosophy of the Infinite* – Rudy Rucker is a prolific, idiosyncratic and always entertaining writer. Science fiction fans will know of him as one of the founding fathers of the cyberpunk genre, but he has written numerous non-fiction books dealing with science, mathematics, computer technology, and philosophy. Rucker is one man who perceives no gap between science and art. Incidentally, his website is well worth a visit.
  - Schacter, Daniel L. *The Seven Sins Of Memory: How the Mind Forgets and Remembers* – Schacter, a Nobel laureate, is another example of a scientist who writes engaging prose that surpasses that of many a 'literary' author. This book is both a wonderfully

clear explanation of how memory works and, incidentally, an excellent practical guide to improving one's memory.

- **Longing For The Infinite: Looking For God And His Cronies**
  - General suggestions and remarks.
    - o It isn't hard to find examples of otherwise brilliant and creative men and women who have written about and defended the most outrageous and ludicrous beliefs in an attempt to find meaning in their lives. The urge to make sense of the world, to find purpose and meaning in life, sometimes leads even the best minds to nonsense. So I've limited this list to two very different books about the all-too-human search for meaning—and two books about the subjects of this chapter's case studies.
  - Recommended reading.
    - o Clark, Ronald W. *Einstein: The Life and Times* – There are numerous Einstein biographies. I only recommend this one, because it is the only one I've read cover to cover—and I thought it excellent. (Einstein scholars might differ in their recommendations, and there are many sources of biographical information on the man to be found on the Internet and in books dealing with a wide range of subjects.)
    - o Dawkins, Richard. *The God Delusion* – Needless to say, Dawkins is one of our most outspoken critics of religion. This brave book pulls no punches and shows no respect. It is book that needed to be written, and we can 'thank the gods' (ahem) that it was written by such an articulate and witty defender of reason in an age where 'tolerance' and 'respect' for dangerous and harmful superstition, because it cloaks itself in the protective mantel of 'religion', is considered polite and right.
    - o Frankl, Viktor. *Man's Search for Meaning* – This book has deep personal significance for me, because it (not medication) was what many, many years ago guided me out of what I'm sure would be considered a "clinical depression". It is a wise book

by a wise man who survived a Nazi concentration camp.

- o Quinones, Ricardo J. *Dante Alighieri* – It is surprising that there are very few detailed biographies of this writer who is considered one the "greats". Online information about his life is easy to find, as are critical essays about his work, but book length biographies seem to be extremely rare. I include this one only on the basis of the reputation of the editors of the series of which it is a part and because I felt I should point the interested reader somewhere.

- **The Conservative Urge: Fading Images**
  - General suggestions and remarks.
    - o Rarely acknowledged is the conservative streak that exists in even the most revolutionary thinkers and artists. Besides my two 'cases' (Pascal and Pound) I've suggested in this list two important books that address this issue. One has to do with conservatism in science (Kuhn) and the other has to do with conservatism in the arts (Johnson). I've also included Shlain's fascinating book on the merging of metaphor in early 20th Century art and science.
  - Recommended reading.
    - o Kuhn, Thomas. *The Structure of Scientific Revolutions* – Like Snow's *Two Cultures*, this is virtually required reading for anyone interested in the central thesis of Agents.
    - o Johnson, Paul. *Art: A New History* – I include this book on my list *not* because I have any admiration for the book or respect for most of the author's opinions. I include it because it is an interesting and articulate example of intellectual conservatism regarding the visual arts. Johnson speaks for the 'silent majority' that reject all the artistic revolutions that marked the last century.
    - o Pascal, Blaise. *Pensées* – This translates as "Thoughts" and was an 'apology' for belief in Christianity, unfinished at the time of death and published posthumously. It contains his famous "Pascal's Wager" putatively 'proving' that betting on belief, on theism, and an afterlife was a no-lose bet. Despite the dubious nature of this argument, it is a

book worth dipping into for the insight it gives into how a brilliant mind can attempt to justify irrational belief.

- o Pound, Ezra. *ABC of Reading* – Pound was very prolific, and it is easy to find a copy of his *Cantos*, or some selected poems, or his famous translations (more like reinventions) of classic poems from the Chinese—all of which are certainly worth reading. And anyone interested in his strange politics can hunt up the "L'America, Roosevelt e le Cause della Guerra Presente" essay or his book *What Is Money For?* I'm including his *ABC* in this list because it highlights the deep aesthetically conservative streak in this most revolutionary of poets.
- o Shlain, Leonard. *Art & Physics: Parallel Visions In Space, Time & Light* – This fascinating book explores how the revolutionary changes in the dominant metaphors in science and art co-evolved and influenced each other.

- **Dangerous Seductions: The Psychic Circus**
  - General suggestions and remarks.
    - o Scepticism is the forgotten virtue. If it replaced faith in the theological trinity of virtues (faith, hope and charity) a lot of harm could be prevented. For anyone interested in cultivating this virtue, the Internet, while a minefield of traps that only scepticism could guide one through, is also a good place to search out the latest intelligent and sceptical evaluation of the many dubious claims to receive wide media coverage. Recommended are James Randi's site and the website for *The Skeptical Inquirer*.
  - Recommended reading.
    - o Eiseley, Loren. *Darwin's Century: Evolution and the Men Who Discovered It* – While Wallace is not the central player in this book, it does present a vivid portrait of the man and the description of the challenges facing serious scientific thinkers in a time when religious belief was assumed to be natural, reasonable and the norm—even among those doing science.
    - o Gardner, Martin. *Science: Good, Bad, and Bogus* – Martin Gardner wrote the "Mathematical Games" column for *Scientific* American for almost thirty years

and has published over sixty books on science and math for the literate layman. In this book he does a great job of debunking paranormal silliness and pseudo-science, while casually educating his reader about how real science works.

o   Shermer, Michael. *Why People Believe Weird Things* – Michael Shermer is the founder of The Skeptics Society. As a former believer in what he calls "a host of weird things" including pyramid power, rolfing, acupuncture, and fundamentalist Christianity, he is empathetic toward the victims of 'belief' while very effectively showing why we are all too easily convinced of the truth of really weird things which objectively make no sense at all.

o   Sagan, Carl and Druyan Ann. *The Demon-Haunted World: Science as a Candle in the Dark* – Another wonderful and entertaining book debunking all the pseudo-science and paranormal nonsense given too much credence in our oddly credulous (since allegedly literate) world. Sagan manages to do this without being abrasive, and so has probably been more effective than many other writers on this topic in getting people to examine and reconsider their casual, naïve beliefs.

AUTHOR'S NOTE

"I believe that literature, like science, is a way of exploring different perspectives; and I believe that the results of these literary explorations, like the results of science, are always inherently tentative. It is for this reason that I choose to call my major works 'hypotheses'. *Secret Agents Present: Looking Through A Glass Darkly* is Hypothesis 17."

ABOUT THE AUTHOR

Ken Stange is the author of 16 books of poetry, fiction, and non-fiction, as well as hundreds of publications in literary and scientific journals. He was the winner of the 2011 Exile/Vanderbilt prize for short fiction, and he is also a visual artist and Professor Emeritus at Nipissing University where he continues to teach "The Psychology of Art" as an online course. His special interest is the relationship of art and science and creativity.